Lecture Notes in Computer Science

Edited by G. Goos and J. Hartmanis

120

Louis B. Rall

Automatic Differentiation: Techniques and Applications

Springer-Verlag
Berlin Heidelberg New York 1981

Author

Louis B. Rall
University of Wisconsin-Madison, Mathematics Research Center
610 Walnut Street, Madison, Wisconsin 53706, USA

AMS Subject Classifications (1980): 68-02, 68 C 20, 65 D 30, 65 G 10,
65 H 10, 65 K 10
CR Subject Classifications (1981): 1.1, 5.1, 5.11, 5.15, 5.16

ISBN 3-540-10861-0 Springer-Verlag Berlin Heidelberg New York
ISBN 0-387-10861-0 Springer-Verlag New York Heidelberg Berlin

Printing and binding: Beltz Offsetdruck, Hemsbach/Bergstr.
2145/3140-543210

PREFACE

This book is based on the notes for a series of lectures given at the Computer Science Department (Datalogisk Institut) of the University of Copenhagen in the second semester of the 1979-80 academic year. The invitation of Dr. Ole Caprani of that institution to present these lectures, as well as his assistance with the course, is gratefully acknowledged. One of the students, Mr. J. W. Owesen, is also thanked for doing the necessary work to make software from the University of Wisconsin-Madison operational at the University of Copenhagen.

The automatic differentiation of functions defined by formulas proceeds by fixed rules, and is conceptually no more difficult than the translation of formulas into code for evaluation. In spite of this, the automatic calculation of derivatives and coefficients of power series has seemed somewhat exotic to numerical analysts, and perhaps too mundane to computer scientists interested in the creation of ever better languages and systems for computation. The purpose of these notes is to fill this intellectual gap, and show that a powerful computational tool can be fashioned without excessive effort.

The choice of topics presented is dictated by personal interest and familiarity with software which actually works, programs which have proved to be durable as well as effective. On the basis of ideas suggested by R. E. Moore, work was begun at the Mathematics Research Center by Allen Reiter in 1964-65 on software for differentiation, generation of Taylor coefficients, and interval arithmetic. This led to interrelated developments in programs for the solution of differential equations, nonlinear systems of equations, numerical integration, interval arithmetic, and a precompiler for the addition of new data types to FORTRAN. (The connection with FORTRAN is one of the reasons for the durability of this software.) This period of activity came to an end in 1977-78 with the departure of Julia Gray, F. Crary, G. Kedem, and J. M. Yohe from the Mathematics Research Center. Significant contributions were made along the way by J. A. Braun, D. Kuba, T. Ladner, T. Szymanski, and H. J. Wertz, among others. The support of the U. S. Army Research Office during the entire period of the development of this software is appreciated.

It is not implied that the subject of these lectures is a closed book; rather, it is an open door for future developments. To this end, each topic has been provided with suggestions for projects ranging from simple exercises to the construction of elaborate computational systems.

The production of these notes was assisted by Carol Gubbins, who did a professional job of preparation of the figures. First and foremost, thanks are due to my wife Fran for untiring patience, support, and help with every step of this project from beginning to end.

Madison, Wisconsin: May, 1981

TABLE OF CONTENTS

CHAPTER I

INTRODUCTION

The utility of computers for evaluating functions defined by formulas has long
been recognized. Given the values of the input variables and parameters (data), a
sequence of arithmetic and other operations will be executed very rapidly to obtain
the desired output values (results). In addition to the values of functions, many
applications of mathematical analysis to various problems require the values of
derivatives of the functions being considered or of the coefficients of their expansions into power series. Since differentiation of functions defined by formulas
is a mechanical process done according to fixed rules, it is highly suitable for
automation along the same lines as function evaluation. The usefulness of digital
computers is increased by the existence of software for automatic differentiation,
since this permits expansion of the scope of mathematical analysis which can be
applied to problems without additional laborious and possibly erroneous hand computation [74].

A brief description will now be given of the organization of material in these
notes. Since differentiation of a function defined by a formula depends on the
translation of that formula into a list of instructions for a sequence of execu-
table operations, a suitable methodology for formula translation will be discussed,
based on the Kantorovich graph [36]. The differentiation of functions represented
in this way will then be considered by two different approaches, one of which processes the list of instructions, and the other uses well-known formulas for the recursive generation of Taylor coefficients [53], [55]. Here, processing the list of
instructions does not mean the use of a high-level "list processing" language such
as LISP [22], [43] for this task, but rather refers to special-purpose software
which transforms the list of instructions for the evaluation of a function into
lists of instructions (or subroutines) for the evaluation of its desired derivatives.

There are also several possible ways in which the evaluation of the derivative
can be implemented. The first one to be described is simply interpretive execution
of the list of instructions defining the derivative to be evaluated. While slow,
this method is suitable for exploratory, interactive computation from a terminal.
A second method of implementation is to compile the list of instructions obtained
for the derivative as a subroutine in an object program. This could be done if the
goal is to obtain an efficient final program to be used often for production computation. The third method to be described is based on the use of a precompiler, such

as AUGMENT [13], [14], [15], [16], which allows the user to declare derivatives or Taylor coefficients to be new data types, and has the effect of allowing the programmer to write in a high level language which includes the operations of differentiation and power series expansion.

Following the discussion of the principles of construction and implementation of software for automatic differentiation in general terms, some specific programs for this purpose will be described. These were developed at the Mathematics Research Center of the University of Wisconsin-Madison, and have been used successfully for a number of years. The programs CODEX [30], [76] and its successor, SUPER-CODEX [89], [90] are of the instruction list processing type, while TAYLOR [78], [80] performs recursive generation of Taylor coefficients. The program TAYLOR-GRADIENT [37], [95] uses the capabilities of AUGMENT to permit the declaration of vectors of first derivatives (gradients) or Taylor coefficients as new data types.

Two programs which make use of the power of automatic differentiation to solve problems in numerical analysis will be presented. One of these is the program NEWTON [26], [39], which is written for the solution of systems of n nonlinear equations in n unknowns numerically by Newton's method, and the analysis of the error of the approximate solutions obtained. This program uses automatic differentiation to calculate the Jacobian matrix of the system of equations to be solved and, in connection with software for interval arithmetic [53], [55], allows the user to apply the theorems of L. V. Kantorovich [35] or R. E. Moore [54], [73] automatically to obtain rigorous error bounds for the difference between the solution actually computed and the true result. The other program, INTE [28], is designed to perform error analysis of numerical integration methods automatically. This program uses the differentiation software to evaluate the mathematical expression for the truncation error term, and interval arithmetic to bound this term and the error resulting from roundoff and inexact coefficients in the integrand in a rigorous manner.

In the final chapter, other differentiation software developed elsewhere, and additional applications will be noted. Brief mention will be made of some programs using automatic differentiation for the solution of differential equations by series expansions, and some other possible applications of differentiation software, such as to constrained and unconstrained optimization problems, will be indicated but not pursued.

The purpose of this book is to present some general principles for construction of software for automatic differentiation of functions defined by formulas, some specific examples of such software, and a few of the many possible applications. As with any other tool, the actual areas of applications are really limited only by the needs and ingenuity of the user. As in the case of automatic function evaluation, it can be expected that automatic differentiation will allow the application of a wider variety of mathematical methods to problems in physical and social sciences, engineering, and other areas, as well as to numerical analysis.

CHAPTER II

FORMULA TRANSLATION

The beginnings of computer science as a subject distinct from the branches of mathematics and electrical engineering concerned with numerical and electronic computation may be traced back to the development of higher-level languages, first assembly languages for organizing the coding for particular models of computers, and then compilers, which can translate programs written in a certain prescribed way, similar to ordinary mathematical and vernacular notation, into code for machines for which the compiler, rather than the user's program, was written. The freedom of expression and the independence of details of specific machines given to programmers by compilers extended the use of mathematical and computational analysis to a horde of problem areas in a manner which may be best described as explosive.

One feature, common to most compilers, is their ability to perform formula translation. In order to program the evaluation of the function $f(x,y)$ given in ordinary mathematical notation by

(2.1) $$f(x,y) = (xy + \sin x + 4)(3y^2 + 6),$$

the user of such a compiler would only need to write an expression of the form

(2.2) $$F = (X*Y + SIN(X) + 4)*(3*Y**2 + 6),$$

or something similar. In (2.2), the values of the variables X, Y and the constants 2,3,4,6 are considered to be data, and the value of F computed from them according to the given formula is called the result of the calculation.

(The form of (2.2) resembles FORTRAN, which, in its various dialects, is the most widely used language for scientific and engineering computation in the U.S.A. In fact, all the software to be described later in these notes is written essentially entirely in FORTRAN. The discussion of principles given here, however, is completely independent of FORTRAN, and applies to any language with similar (or greater) capabilities.)

The job of the formula translation segment of the compiler being used is to accept an expression such as (2.2) as data, assign storage locations to the various quantities appearing in the computation, and finally to produce a sequence of machine instructions which will give the numerical value of F when executed. The exact method by which this is done by a given compiler is not important at the present stage of the discussion. The main point is, however, that each compiler translates formulas by the application of a fixed set of rules which apply to all formulas which are "legal" in the computer language being used.

1. Function evaluation. Attention will be confined here to functions which can be evaluated by performing a sequence of arithmetic operations and calls to subroutines which compute the values of what will be called library functions. Examples of arithmetic operations and their associated symbolism would be addition +, subtraction -, multiplication *, division /, and exponentiation **. The set of library functions would usually include the square root, the sine, cosine, tangent, and their inverses, natural and common logarithms, exponential functions with bases e and 10, and perhaps some special functions useful in a particular problem area, such as hyperbolic functions and inverses, Bessel functions, and Legendre polynomials. The distinction between arithmetic operations and library functions is somewhat arbitrary, as arithmetic operations can (and frequently will) be performed by a subroutine, particularly on mini and microcomputers, and when some nonstandard type of arithmetic, such as interval arithmetic [93], [94], is being used. Furthermore, operations other than the ones listed above could be considered. For example, one may prefer to limit the set of arithmetic operations to addition (+), subtraction (-), multiplication (*), and reciprocation (1/ or **-1) [37], [41], [42], [44], [95]. In this case, division would be done by reciprocation of the divisor, followed by multiplication by the dividend, and exponentiation would also be a library function. For the present purpose, it will be convenient, but not essential, to consider the set of arithmetic operations to consist of +, -, *, /, **, and the library functions to be functions of a single variable, including the elementary functions (square root, sine, cosine, ...) and any special functions (Bessel functions, Legendre polynomials, ...) appropriate to the problem at hand. Functions which can be evaluated by a sequence of arithmetic operations and library functions using input or previously computed values as data will be called codeable functions. Among the codeable functions, consideration will be given for the most part to functions which can be expressed simply by formulas similar to (2.2) in the computer language being used.

By an analysis of the formula for the function considered into a sequence of arithmetic operations and calls to library subroutines (subroutines for library functions), one obtains an equivalent representation of the function as a code list. In order to illustrate this process, the function (2.1) can be represented by the sequence of instructions indicated in the following code list:

$$
\begin{aligned}
T1 &= X*Y \\
T2 &= \text{SIN}(X) \\
T3 &= T1 + T2 \\
T4 &= T3 + 4 \\
T5 &= Y**2 \\
T6 &= 3*T5 \\
T7 &= T6 + 6 \\
F &= T4*T7
\end{aligned}
$$

(2.3)

Note that the code list is itself a sequence of statements in the same language in which the formula (2.2) is written, and hence could be translated by the same compiler into machine language for execution. Each line in the list (2.3) contains only a _binary_ operation (i.e., an arithmetic operation) on two input data or previously computed values, or a _unary_ operation on a single piece of data or previously computed value, illustrated in this example by the call to the library subroutine for the sine function, in the terminology of Reiter [30], [76]. Since formulas of the form (2.2) will be considered to be input data for a program (called the _coder_) which produces lists (2.3) as output, the constants 2, 3, 4, 6 appearing in the formula will be considered to be input data for the calculation of the value of F, in addition to the values of the variables X, Y. In what follows, the discussion will be based on the equivalence of the _function code list_ (2.3) to the _formula_ (2.2) for F.

It should be mentioned that what are called codeable functions here are termed _factorable functions_ by McCormick [41], [42], [44] and Kedem [37], [95]. Kedem [95] (see also [37]) gives the following definition of a factorable function.

Let f be a map from R^n into R^m. Superscripts will be used to denote components; $x = (x^1, x^2, \ldots, x^n) \in R^n$, $f(x) = (f^1(x), f^2(x), \ldots, f^m(x)) \in R^m$, subscripts will index functions f_1, f_2, ... The set of computable (library) functions and operations is denoted by Λ. A function f is said to be factorable if and only if there exists a _finite_ sequence of functions f_1, f_2, ..., f_k: $D \subset R^n \to R$ that satisfy the following conditions:

1) $f_1(x) = x^1$, $f_2(x) = x^2$, ..., $f_n(x) = x^n$;

2) $f_{k-m+1}(x) = f^1(x^1, \ldots, x^n)$, ... , $f_k(x) = f^m(x^1, \ldots, x^n)$;

3) for $i = n + 1$, ..., k, either there exists a $g \in \Lambda$ such that

$$f_i(x) = g(f_{j_1}(x), \ldots, f_{j_s}(x)), \quad j_1, j_2, \ldots, j_s < i,$$

or

$$f_i(x) \equiv c_i, \quad c_i \in R \text{ a constant.}$$

This definition is considerably more general than needed in the following discussion, as it deals with the evaluation of m functions $f^1(x)$, $f^2(x)$, ..., $f^m(x)$ of n input variables x^1, x^2, ..., x^n, and the set Λ of operations and library functions is allowed to include functions of an unspecified (but finite) number of variables. Kedem calls f_1, f_2, ..., f_k a _basic sequence_, and says that this sequence is a _basic representation_ of f [37], [95]. The description of a codeable function given above can be fit into the framework of this definition by taking m = 1, so that f(x) = $f^1(x)$, and by restricting the library functions g to be functions of one variable (the unary library routines), or one of the functions of two variables allowed as a binary arithmetic operation. The basic sequence f_1, f_2, ..., f_k is then closely related to the _code list_ for f, which will be defined precisely in Chapter 3; for the present, the structure of a code list for a function is illustrated adequately by the example (2.3).

A code list for a codeable function, however, differs from a basic representation of the function in that the basic sequence contains the input variables x^1, x^2, x^n and the constants C_1, C_2, ... involved in the calculation explicitly, while these are treated as <u>data</u> in the code list, and their values are assumed to be entered in a <u>data base</u> outside (but accessible to) the code list of instructions. It is also considered advisable to reject the terminology "factorable function" on the grounds that this has a well-established connotation in ordinary algebra as a function which can be expressed as the multiplicative product of simpler factors.

2. <u>The Kantorovich graph of a codeable function</u>. The importance of the code list in the evaluation and further analysis of a codeable function is evident from examination of a typical example, such as (2.3). Each line of (2.3) presents a very simple problem in differentiation and, as will be seen later, the methods to be presented for automatic differentiation of functions depend on having a code list for the evaluation of the function. Consequently, the <u>coder</u>, which produces the code list from the given formula for the function in question, is the key part of the software for differentiation. The construction of a coder, however, follows the principles for formula translation, which are well-known by now and used in the design of most compilers. A simple type of coder, the one used in CODEX [30], [76] will be described later in Chapter 5. It is fair to say that formula translation presents a somewhat greater challenge than differentiation, in that most people apply rules for formula evaluation subconsciously, so that different people would perform operations on the same formula in a different sequence, and even the same person might do so when evaluating the given function at a different time. In writing a program for formula translation, however, all the rules used must be made explicit, they must apply to all correctly written formulas, and they must always produce the same sequence of instructions when applied to the same formula. In other words, automatic formula translation has to be done in a conscious, rather than an unconscious, manner.

Since there is not in general a unique code list for a given codeable function f, it is helpful to have a method for the visualization of the evaluation of the function from which one or more valid code lists can be derived. A device of this type is furnished by the <u>Kantorovich graph</u> of the calculation [36]. An example of such a graph, again for the calculation of f = f(x,y) defined by (2.1), is shown in Figure 2.1 on the following page. This graph is seen, first of all, to be a <u>directed</u> graph, somewhat similar to a flow chart, in which information (in this case, numerical values) is transmitted along the edges in the direction indicated by the arrows, generally from the top downward in the given depiction. Secondly, although the nodes of the graph are labeled to correspond to (2.3), many possible code lists for f may be read directly from the graph.

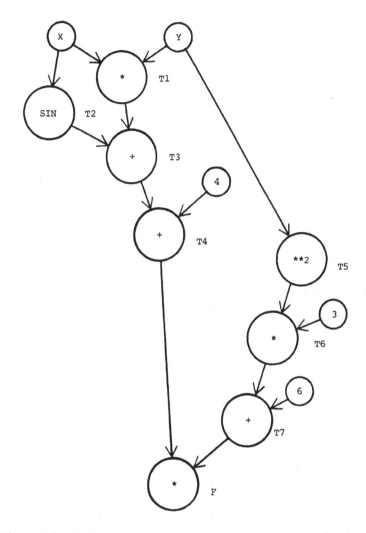

Figure 2.1. A Kantorovich Graph of the Calculation of f(x,y).

For example,

$$T5 = Y**2$$
$$T6 = 3*T5$$
$$T7 = T6 + 6$$
$$T1 = X*Y$$
$$T2 = SIN(X)$$
$$T3 = T1 + T2$$
$$T4 = T3 + 4$$
$$F = T4*T7$$

(2.4)

is also a code list for f, and corresponds to starting down the right side of the
graph and going as far as possible downward before shifting to the left side. It

is possible to obtain another code list by regarding the edges of the graph as being equal in length, and going down the graph level by level. This gives the list

(2.5)

$$
\begin{aligned}
T1 &= X*Y \\
T2 &= SIN(X) \\
T5 &= Y**2 \\
T3 &= T1 + T2 \\
T6 &= 3*T5 \\
T4 &= T3 + 4 \\
T7 &= T6 + 6 \\
F &= T4*T7
\end{aligned}
$$

By renumbering lines in (2.4) and (2.5) (i.e., relabeling the nodes of the graph in Figure 2.1), code lists of the form (2.3) with T1, T2, ..., T7, F in sequential order can be obtained.

A code list constructed according to the procedure used to obtain (2.5) may be useful in connection with __parallel computation__. In this case, as many lines of the code list as possible would be assigned to separate components of a __multiprocessor__ for simultaneous execution. In a parallel environment, the evaluation of f(x,y) could be done in four steps:

(2.6)

$1°.$ T1 = X*Y, T2 = SIN(X), T5 = Y**2;

$2°.$ T3 = T1 + T2, T6 = 3*T5;

$3°.$ T4 = T3 + 4, T7 = T6 + 6;

$4°$ F = T4*T7.

This last example is given to indicate the wide usefulness of the Kantorovich graph as a programming tool. The methods of differentiation to be discussed below are applicable to the parallel computational scheme given in (2.6) without conceptual modification. However, the presentation of automatic differentiation given below will be in the context of ordinary sequential computation, in which software for implementation actually exists. In particular, the software to be described in Chapter 5 produces code lists by a left-to-right analysis of formulas of the form (2.2) in much the same fashion as an ordinary FORTRAN compiler operates.

CHAPTER III

FORMULA DIFFERENTIATION

In contrast to the rules for formula evaluation, the rules for differentiation
are known explicitly from elementary calculus, and are applied in a very conscious
fashion by most people. Even at its best, however, differentiation of a formula is
a dull, uninteresting task which, like washing dishes, is probably best left to a
machine. In spite of the mechanical nature of the differentiation process and the
importance of derivatives in mathematical analysis since the time of Newton and
Leibniz, the use of computer software for automatic differentiation is not as wide-
spread as one might expect. In this chapter, some of the principles for the construction of programs for the differentiation of functions defined by formulas will
be described in general terms.

1. Rules for differentiation. The basic idea behind the automation of differentiation is very simple: Once a code list has been obtained for the function
considered, then the rules of elementary calculus can be applied to it line by line
to give a list of instructions for evaluation of the derivative. Thus, the coder,
which is the program which translates the formula for the function into the code
list, is the key piece of software in the process of differentiation as well as
evaluation of the function. Of course, a codeable function is not necessarily differentiable. By a differentiable function will be meant a codeable function defined
on a set Λ of library functions such that if its code list contains a function $g \in \Lambda$,
then the derivative g' of g is a codeable function defined on Λ. This requirement,
which is satisfied, for example, if Λ consists of the arithmetic operations and the
elementary algebraic, trigonometric, logarithmic, exponential, and hyperbolic functions, implies that the derivative f' of f is a codeable function over the set Λ of
library functions. It should be noted that to say a function is differentiable does
not mean that its derivative (or even the function itself) can be evaluated for all
values of the input variables. In the execution of the sequence of instructions in
the code list, attempts to divide by zero or evaluate logarithms of nonpositive numbers, for example, should result in the usual type of error indication.

It will be useful to devote a certain amount of attention to the meaning of the
term "derivative" as it will be used in these notes. In most applications, what is
wanted is one or more partial derivatives of a function with respect to some of the
variables entering into the formula for it. For example, for the function defined
by (2.1), one may wish to obtain

(3.1)
$$\frac{\partial f(x,y)}{\partial x} = (y + \cos x)(3y^2 + 6),$$

or

(3.2)
$$\frac{\partial f(x,y)}{\partial y} = x(3y^2 + 6) + 6y(xy + \sin x + 4)$$

$$= 9xy^2 + 6y \sin x + 6x + 24y.$$

These formulas are obtained by the rules for differentiation of functions of a single variable, treating all other variables entering into the formula for the function as constants. Thus, in case all variables are independent, the partial derivative $\partial f/\partial x$ would have the interpretation of giving the instantaneous rate of change of the value of the function f with respect to x, a very useful concept in physics, economics, and elsewhere. If, however, the variable y is itself a function of x and hence not an independent variable, then f(x,y) is actually a function of x alone, and the desired rate of change is an ordinary derivative, given by

(3.3)
$$\frac{df}{dx} = \frac{\partial f}{\partial x} + \frac{\partial f}{\partial y} \cdot \frac{dy}{dx}$$

at the current values of x,y. In a more general case, suppose that f and y also depend on variables u, v, w, ... which _are_ independent of x. Then, the rate of change of f with respect to x is the so-called _semi-total_ derivative of f with respect to x, which is denoted by $Df/\partial x$ [83] and given by the formula

(3.4)
$$\frac{Df}{\partial x} = \frac{\partial f}{\partial x} + \frac{\partial f}{\partial y} \cdot \frac{\partial y}{\partial x} ,$$

at the current values of the variables. The extension of these concepts and formulas to functions of n variables is straightforward, and may be found in standard texts on multivariate calculus (for example, [83]).

It should be mentioned that while the distinction between the _formal_ partial derivative $\partial f/\partial x$ and the rate of change of f with respect to x given by the semi-total derivative $Df/\partial x$ is fairly simple, failure to observe it can lead to conceptual and computational errors. Of course, if the variable y is independent of x, then $\partial y/\partial x = 0$ in (3.4), and the two derivatives coincide. In complicated programs, the formula defining y as a function of x and perhaps other variables may be remote from the formula for the function f to be differentiated, so some care may have to be taken to compute the derivative actually desired correctly. This occurs also in fairly simple programs, since many people prefer to code a complicated formula in several lines, rather than in a single line. For example, a formula such as

(3.5)
$$G = X**2 + EXP(X*(Y + X*Z)/Y) + Y + X*Z$$

might be coded as

(3.6)
```
U = Y + X*Z
V = (X*U)/Y
G = X**2 + EXP(V) + U .
```

A formal partial differentiation of G as defined in the last line of (3.6) with respect to X gives simply 2*X; to obtain the partial derivative of the function G given in (3.5) with respect to X from the sequence (3.6) (which evaluates G correctly) requires differentiation of U and V and application of an extended form of (3.4). Although the code lists to be differentiated by the software to be described later are simpler in structure than (3.6), the same principle applies, and it must also be taken into account that formulas in other segments of the program, and hence other code lists, may define some of the variables appearing in the list in terms of the variables with respect to which derivatives are desired.

Thus, in order to obtain derivatives of differentiable functions automatically, repeated applications of the chain rule of elementary calculus will usually be required. For composite functions $f(x) = g(h(x))$ of one variable, the chain rule gives

(3.7) $$f'(x) = g'(h(x)) \cdot h'(x).$$

In terms of the operation of composition of functions, denoted by $\circ$, the function f could be written $f = g \circ h$, and its value at the point x is then given by $f(x) = (g \circ h)(x)$. The formula (3.7) for the derivative with respect to x becomes

(3.8) $$f'(x) = (g' \circ h)(x) * h'(x),$$

where $*$ denotes ordinary multiplication. It follows from (3.7) and (3.8) that the evaluation of the derivative f' of f at x to obtain $f'(x)$ requires the differentiation of g and h to get the functions g' and h', the evaluation of h' at x to obtain the value $h'(x)$, and then either composition of g' and h, followed by evaluation of the composite function $g' \circ h$ at x, or the evaluation of g' at the point $h(x)$ to obtain the multiplier $g'(h(x)) = (g' \circ h)(x)$ of $h'(x)$. The same considerations apply to a function f which is the composition of a finite number n of functions $g_1, g_2, \ldots, g_n$, that is,

(3.9) $$f = g_1 \circ g_2 \circ \ldots \circ g_{n-1} \circ g_n .$$

At a point x, the derivative $f'(x)$ of f is given by

(3.10) $$f'(x) = (g_1' \circ g_2 \circ \ldots \circ g_{n-1} \circ g_n)(x) * (g_2' \circ g_3 \circ \ldots \circ g_{n-1} \circ g_n)(x) * \ldots * (g_{n-1}' \circ g_n)(x) *$$
$$* g_n'(x) ,$$

by repeated application of the chain rule. Thus, in order to evaluate the derivative of the composite function f given by (3.9), the n derivatives $g_1', g_2', \ldots, g_{n-1}', g_n'$ are needed, and the values of the n factors on the right side of (3.10). The representation of f as a composite function by (3.9) is closely related to the idea of the basic representation of a function by a basic sequence, and thus to its representation in terms of a code list. From (3.9), if one sets

$$f_0 = x$$
$$f_1 = g_n(f_0)$$
(3.11) $$f_2 = g_{n-1}(f_1)$$

$$\cdots \quad \cdots \quad \cdots$$

$$f_{n-1} = g_2(f_{n-2})$$

$$f_n = g_1(f_{n-1}) \ ,$$

then it is evident that the calculation of the sequence of values of f_1, f_2, ..., f_n yields the value $f(x) = f_n$ of the function f at x. Furthermore, given this sequence and the derivatives g'_i, $i = 1,2,\ldots,n$, one may compute the values

$$f'_0 = 1$$

$$f'_1 = g'_n(f_0) * f'_0$$

$$f'_2 = g'_{n-1}(f_1) * f'_1$$

(3.12)

$$\cdots \quad \cdots \quad \cdots \quad \cdots$$

$$f'_{n-1} = g'_2(f_{n-2}) * f'_{n-2}$$

$$f'_n = g'_1(f_{n-1}) * f'_{n-1} \ .$$

The chain rule now asserts that the value of the derivative f' of f at x is $f'(x) = f'_n$, that is

(3.13)
$$f'(x) = \prod_{i=1}^{n} g'_i(f_{n-i}) * f'_0$$

$$= g'_1(f_{n-1}) * g'_2(f_{n-2}) * \cdots * g'_{n-1}(f_1) * g'_n(f_0) * f'_0 \ ,$$

as required by (3.10). The generalization of these ideas to functions of several variables will be explained later, and will be based on extensions of the definitions of derivatives and the operation denoted above by *, which is ordinary multiplication in the scalar (one variable) case. Before going on to that, some observations in connection with the above material are in order.

First of all, the connection between the composite function f represented by (3.9) and its evaluation at x by means of the sequence (3.11) is obvious. It is also clear that given the list (3.11), the representation (3.9) of f can be reconstructed. Also, if the functions g_i, $i = 1,2,\ldots,n$ are library functions, then (3.11) is a code list for f. Furthermore, if the derivatives g'_i, $i = 1,2,\ldots,n$, are also library functions, then f is differentiable, and (3.12) is almost (but not quite) a code list for the derivative f'. In order to convert (3.12) into a code list, all that is required is to replace each line (except the first) by two lines, that is, $f'_k \leftarrow d'_{2k-1}$, d'_{2k}, $k = 1,2,\ldots,n$, where

(3.14)
$$d'_{2k-1} = g'_{n-k+1}(f_{k-1})$$

$$d'_{2k} = d'_{2k-1} * d'_{2k-2} \ ,$$

with $d'_0 = f'_0 = 1$. It is easy to see that (3.11) followed by the sequence d'_1, d'_2, ..., d'_{2n} obtained in this way is a code list for the evaluation of the derivative f', and that $f'(x) = d'_{2n}$.

The key to the representation of the function f by the list (3.11) is, of course, its representation (3.9) as the compostition of the functions g_1, g_2, ..., g_n, which has the form of a "factorization" of f with respect to the operation ° of composition, rather than multiplication, over a set of functions Λ which includes the g_i, i = 1,2,...,n. It was this observation which gave rise to the terminology "factorable function" in case Λ is the set of library functions. In the present context of codeable functions, it may be noted that if each function g_i, i = 1,2,..., n, is codeable (which includes the possibility that a given g_i is simply a library function), then f is codeable, and a code list for f can be obtained by inserting the code list for each such g_i into the appropriate place in the list (3.11). Similarly, if each g_i is differentiable, then f is differentiable, and a code list for f' may be obtained by replacing d'_{2k-1} in (3.14) by the code list for evaluation of $g'_{n-k+1}(f_{k-1})$.

The above discussion is not quite satisfactory for our purposes, as it is concerned entirely with representations of functions in terms of library functions of one variable only, and needs to be completed for the case of codeable functions by discussion of code lists which contain the arithmetic operations +, -, *, /, **, which are functions of two variables in this context, albeit very simple ones. The case of one variable, however, is instructive in several respects, which apply also to the more general case. First of all, note that while the list (3.12) for the evaluation of f'(x) does not require the value of $f_n(x)$ from the list (3.11), it does require the previous values f_{n-1}, f_{n-2}, ..., f_1, and, of course, $f_0 = x$. Thus, one would ordinarily expect to go through the process of evaluation of f(x) before starting the evaluation of f'(x), at least in ordinary sequential computation. (An examination of (3.11) and (3.12) reveals the possibility of simultaneous evaluation of f_n and f'_n by a sufficiently capable parallel processor.) Even in ordinary differentiation, the value of f(x) may be convenient to use directly in the evaluation of f'(x). The example which comes to mind immediately is, of course,

(3.15) $f(x) = e^x = \exp(x)$, $f'(x) = f(x)$.

Another example is based on another approach to the differentiation of x^n than the formula learned early in the study of calculus:

(3.16) $f(x) = x^n$, $f'(x) = nx^{n-1}$.

In many cases, it is more efficient in computation to use the alternative formulation

(3.17) $f(x) = x^n$, $f'(x) = nf(x)/x$,

at least if a certain amount of care is exercised. For example, if n > 1 and x ≠ 0, then the use of (3.17) presents no problem, and a test for x = 0 could return the values f(0) = f'(0) = 0 if satisfied. For n = 1, the value f'(x) ≡ 1 is a constant, and should be treated as such in any subsequent differentiations. If n < 1, then

(3.17) would again ordinarily be preferable to (3.16) for $|x| \geq 1$, but questions of numerical accuracy should be considered in case $0 < |x| < 1$, to choose between the division in (3.17) and the use of a logarithm-antilogarithm subroutine to obtain the value of x^{n-1} needed in (3.16). For $0 < n < 1$, an attempt to evaluate $f'(0)$ should lead to an error condition as in the case of $f(0)$ for $n < 0$, and, finally, for $n = 0$, one has the constant values $f(x) \equiv 1$, $f'(x) \equiv 0$. Details of the differentiation of $f(x) = x^n$ will be discussed more fully in Chapter 4, §3.

Other elementary and special functions can also be differentiated by formulas alternative to the ones commonly taught in calculus, which make use of the values of the functions themselves. For example, for the basic trigonometric functions, one has:

$$f(x) = \sin x, \quad f'(x) = \cos x = \sqrt{1 - [f(x)]^2}\,,$$

(3.18)
$$g(x) = \cos x, \quad g'(x) = -\sin x = -\sqrt{1 - [g(x)]^2}\,,$$

$$h(x) = \tan x, \quad h'(x) = \sec^2 x = 1 + [h(x)]^2.$$

This discussion of differentiation formulas is not intended to be exhaustive, but simply to suggest that alternative methods for differentiation to the ones ordinarily used in hand calculation are possible, and may be preferable in certain circumstances for automatic computation of derivatives. This possibility should be kept in mind when designing software for analytic differentiation.

Another result which follows immediately from consideration of the simple case that (3.11) is a code list is that the code list for the derivative $f'(x)$ obtained by applying the transformations (3.14) to the list (3.12) is twice as long as the list for $f(x)$. Since all the terms in (3.11) except perhaps the last are required for evaluation of the sequence d_0', d_1', d_2', ..., d_{n-1}', $d_{2n}' = f'(x)$, one can conclude that essentially $(1 + 2)n = 3n$ lines are required in code lists for the evaluation of the first derivative of a function. The fact that differentiation appears to triple the number of lines in the list may be discouraging at first sight, but this really has little bearing on what occurs in practice. First of all, many important applications require only first or second derivatives, so the lists, while long, can be stored without undue difficulty. Secondly, as will be discussed later, there are special techniques which apply to series expansion, which is an application which requires derivatives up to some fairly high order. Thirdly, the result given above is an _estimate_, and may not actually be attained in a given differentiation; for example, constants and more generally polynomials in the variable of interest will eventually vanish in the differentiation process, and there is thus the possibility that the corresponding code lists will even decrease in length, rather than increasing indefinitely. An example of this will be given in the following section.

2. _Differentiation of code lists_. In order to extend the ideas in the previous section to the types of functions encountered in actual computation, all that is basically required is to allow the functions g_k, $k = 1, 2, ..., n$, in the sequence

(3.9) for the representation of f to be functions of several variables. Regarding the point x as a single number, the function g_k could be allowed to be a function of up to $n - k + 1$ variables, so that the ith line in the list (3.11) could be written in general as

$$(3.19) \qquad f_i = g_{n-i+1}(f_{i-1}, f_{i-2}, \ldots, f_1, f_0),$$

$i = 1, 2, \ldots, n$, where, of course, $f_0 = x$. This degree of generality is not necessary for the discussion of code lists, nor is the possibility of allowing the variable x to be a ν-dimensional vector $x = (x^1, x^2, \ldots, x^\nu)$ as was done by Kedem [37], [95]. All that is required is to augment the set Λ of library functions by the functions of two variables $g(u,v)$ corresponding to the five $\underline{binary}$ arithmetic operations (+, -, *, /, **) allowed in the definition of a codeable function. (It should also be pointed out that the arithmetic operations also define unary operations in case one argument is a constant, at least with respect to the variable of interest, or when both arguments are the same. Of course, if both arguments are constant, then the arithmetic operation simply defines a constant.)

Supposing now that the code list for f contains a line of the form

$$(3.20) \qquad f_k = g(f_i, f_j), \quad i, j < k,$$

and g is a differentiable function, then the sequence for the evaluation of f' will have the term

$$(3.21) \qquad f'_k = g'_1(f_i, f_j) * f'_i + g'_2(f_i, f_j) * f'_j,$$

where

$$(3.22) \qquad g'_1(u,v) = \frac{\partial g(u,v)}{\partial u}, \quad g'_2(u,v) = \frac{\partial g(u,v)}{\partial v},$$

are the (formal) partial derivatives of g with respect to its first and second arguments, evaluated at the indicated point (u,v). If g'_1 and g'_2 are library functions, then f is differentiable, and a code list for f' can be constructed by transformations of the form (3.14) and replacement of terms of the form (3.21) by the subsequence $d'_{k,1}, d'_{k,2}, d'_{k,3}, d'_{k,4}, f'_k$, where

$$(3.23) \qquad \begin{aligned} d'_{k,1} &= g'_1(f_i, f_j) \\ d'_{k,2} &= d'_{k,1} * f'_i \\ d'_{k,3} &= g'_2(f_i, f_j) \\ d'_{k,4} &= d'_{k,3} * f'_j \\ f'_k &= d'_{k,2} + d'_{k,4} \end{aligned}$$

to obtain the corresponding code list. Thus, allowing functions of two variables will expand the code list by a larger factor (in this case, 5) than given before for the production of the code list for the derivative in the case of functions of a single variable only. If g_1' and g_2' are not library functions, but are codeable, then the insertion of the corresponding code lists in place of $d_{k,1}'$ and $d_{k,3}'$ in (3.23) will yield the segment of the code list for f' indicated by f_k' in (3.21).

For most of the arithmetic operations, however, the situation is not as complicated as might be expected from (3.23). For addition and subtraction,

(3.24)
$$f_k = f_i \pm f_j, \quad i,j < k,$$

one has

(3.25)
$$f_k' = f_i' \pm f_j',$$

which does not increase the number of lines in the code list for f'. In the case of multiplication,

(3.26)
$$f_k = f_i * f_j, \quad i,j < k,$$

a formula for f_k' is

(3.27)
$$f_k' = f_j * f_i' + f_i * f_j',$$

to which corresponds the three-line code list

(3.28)
$$d_{k,1}' = f_j * f_i'$$
$$d_{k,2}' = f_i * f_j'$$
$$f_k' = d_{k,1}' + d_{k,2}'.$$

In the case of division,

(3.29)
$$f_k = f_i / f_j, \quad i,j < k,$$

a straightforward application of (3.21) gives the formula

(3.30)
$$f_k' = (1/f_j) * f_i' + (-f_i/f_j **2) * f_j',$$

which, for computational purposes, has the more convenient expression

(3.31)
$$f_k' = (f_i' - f_i * f_j'/f_j)/f_j,$$

which avoids exponentiation. (One could also replace $f_j **2$ in (3.30) by $f_j * f_j$ to the same end.) A code list corresponding to (3.31) is

(3.32)
$$d_{k,1}' = f_i * f_j'$$
$$d_{k,2}' = d_{k,1}'/f_j$$
$$d_{k,3}' = f_i' - d_{k,2}'$$
$$f_k' = d_{k,3}'/f_j,$$

a four-line list. Finally, for exponentiation in the general case, one may write

(3.33)
$$f_k = f_i ** f_j = \exp(f_j * \log(f_i))$$

in terms of the natural exponential and logarithmic functions with the base e = 2.71828... . Differentiation of (3.33) gives the formulas

$$(3.34) \qquad f'_k = \exp(f_j * \log(f_i)) * (f'_j * \log(f_i) + f_j * f'_i / f_i)$$

$$= (f_i ** f_j) * (f'_j * \log(f_i) + f_j * f'_i / f_i)$$

$$= f_k * (f'_j * \log(f_i) + f_j * f'_i / f_i) \ ,$$

and the corresponding code list

$$d'_{k,1} = \log(f_i)$$

$$d'_{k,2} = f'_j * d'_{k,1}$$

$$(3.35) \qquad d'_{k,3} = f_j * f'_i$$

$$d'_{k,4} = d'_{k,3} / f_i$$

$$d'_{k,5} = d'_{k,2} + d'_{k,4}$$

$$f'_k = f_k * d'_{k,5} \ ,$$

which requires six lines, since the partial derivatives of the function defined by (3.33) are not both library functions, the second being a codeable function.

The above analysis is adequate for the computation of ordinary derivatives of functions of a single variable, or formal first partial derivatives if all arguments appearing in the code list except the variable of interest (denoted by f_0 = x above) and names of previous lines are regarded as constants. In most cases, however, one will deal with expressions containing several variables X, Y, Z, ..., one or more parameters P1, P2, ..., and constants 0, 1, -3.5, ..., and so on. By a parameter is understood a symbol which denotes a constant for the purpose of differentiation, but which may be set to different values in successive runs of the same program. (Parameters are the "variable constants" so dear to the hobgoblins of logical consistency.) If necessary to distinguish them from parameters, numbers such as 0, 1, -3.5, ... will be called literal (or numerical) constants. Generally speaking, it is up to the user to determine which symbols represent variables and which represent parameters in any given formula or program.

The formalism of differentiation will now be extended to functions of any number of variables which are represented by code lists satisfying the following restriction: Each argument of a library function or arithmetic operation in the code list is a variable, parameter, constant, or label (name) of a previous line in the code list (for example, the formula for f_5 could involve f_2 and f_4).

One way to obtain partial derivatives of a function defined by a formula with respect to any variable entering into the formula, and to obtain semi-total derivatives of functions defined by several formulas, is by the use of differentials [71], [74]. For a function of a single variable, f = g(v), the differential df is defined by

(3.36) $$df = g'(v)*dv$$

in terms of the differential dv, and, for f = g(u,v), a function of two variables, one has

(3.37) $$df = g'_1(u,v)*du + g'_2(u,v)*dv \ ,$$

where $g'_1(u,v)$ and $g'_2(u,v)$ are the partial derivatives appearing in (3.22), and the differentials du, dv in (3.36) and (3.37) may be considered to be new variables, if not otherwise defined. It follows that formulas and code lists for differentials of library functions and arithmetic operations may be obtained from the ones given above for derivatives simply by replacing f'_i, f'_j, f'_k by the corresponding differentials df_i, df_j, df_k. In an actual code list, this would amount to replacing, for example, the line

(3.38) $$U2 = SIN(U1)$$

by the code list

(3.39)
$$V1 = COS(U1)$$
$$DU2 = V1*DU1 \ .$$

Application of this process to the code list (2.3) for evaluation of the function (2.1) defined by the formula (2.2) yields the code list

$$V1 = Y*DX$$
$$V2 = X*DY$$
$$DT1 = V1 + V2$$
$$V3 = COS(X)$$
$$DT2 = V3*DX$$
$$DT3 = DT1 + DT2$$
$$DT4 = DT3$$

(3.40)
$$V4 = Y**1$$
$$V5 = 2*V4$$
$$DT5 = V5*DY$$
$$DT6 = 3*DT5$$
$$DT7 = DT6$$
$$V6 = T4*DT7$$
$$V7 = T7*DT4$$
$$DF = V6 + V7$$

A code list for $\partial f/\partial x$, for example, can be obtained from (3.40) by setting DX = 1, DY = 0, and eliminating trivialities (multiplications by one or zero, exponentiation to the first or zero power, addition of zero, identical lines, references to single pieces of data (variables, parameters, or constants)) to obtain

(3.41)
$$DXT2 = COS(X)$$
$$DXT4 = Y + DXT2$$
$$DXF = T7*DXT4 .$$

This follows immediately from the formula (3.37). Similarly, setting DX = 0, DY = 1, and performing similar simplifications will result in a code list for $\partial f/\partial y$. For convenience, the formulas and code lists for the library functions and arithmetic operations and their differentials which were actually used in the construction of (3.40) are collected below in Table III.1 on page 30.

Of course, if one is only interested in the <u>values</u> of the partial derivatives $f'_x(x,y) = f'_1(x,y)$ and $f'_y(x,y) = f'_2(x,y)$ at the current values of x,y, then these can be obtained by executing the sequence of instructions (3.40), first with DX = 1, DY = 0, and then with DX = 0, DY = 1. The code list (3.41), however, includes only the variables in the original list (2.3), and may thus be evaluated with the same input data. Furthermore, it may be differentiated by the same process by which it was obtained to yield similar code lists for higher partial derivatives $\partial^2 f/\partial x^2$, $\partial^2 f/\partial x\partial y$, $\partial^2 f/\partial y^2$, $\partial^3 f/\partial x^3$, ..., and so on. There are additional possibilities for simplification if one is only interested in computing some of the derivatives of the original function, and possibly not the function itself or certain intermediate derivatives. This is illustrated for the above example in Figure 3.1 on the following page, which shows the Kantorovich graph of the calculation of both f(x,y) and $f'_x(x,y)$, based on Figure 2.1 (page 7). In the graph in Figure 3.1, the nodes indicated by squares may be eliminated to obtain the simpler graph for the calculation of $f'_x(x,y)$ only, shown in Figure 3.2 on page 21.

This example was chosen deliberately to allay fears of exponential increase in the length of code lists for derivatives in all cases. It is not misleading, due to the fact that polynomial elements are not uncommon in nonlinear formulas. Mathematical modeling of real phenomena usually begins with linear equations, and polynomials are one of the usual first steps in the direction of more accurate formulas which incorporate nonlinear features.

The use of differentials also applies in general to functions of ν variables,

(3.42)
$$f(x) = f(x^1, x^2, \ldots, x^\nu),$$

where $x = (x^1, x^2, \ldots, x^\nu) \in R^\nu$. The <u>differential</u> df of f at x, if it exists, is given by

(3.43)
$$df = \frac{\partial f(x)}{\partial x^1} dx^1 + \frac{\partial f(x)}{\partial x^2} dx^2 + \ldots + \frac{\partial f(x)}{\partial x^\nu} dx^\nu .$$

(This expression is usually called the <u>total</u> differential of f [83].)

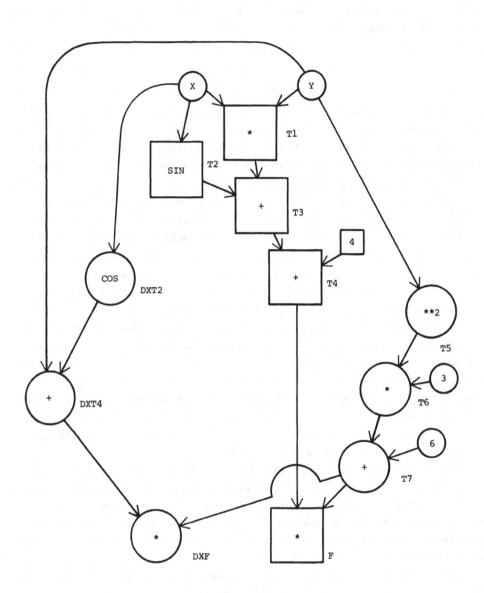

Figure 3.1. A Kantorovich Graph of the Calculation

of $f(x,y)$ and $f'_x(x,y)$.

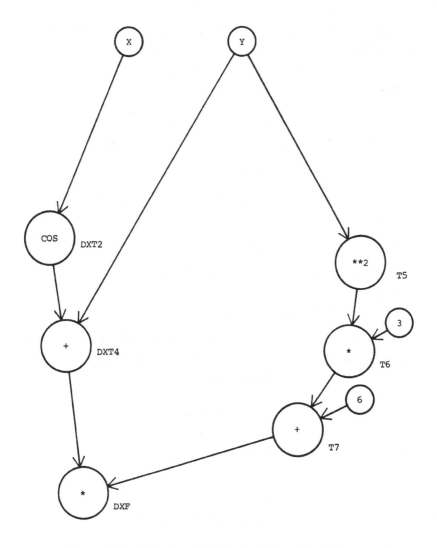

Figure 3.2. A Kantorovich Graph of the Calculation

of $f'_x(x,y)$ Only.

The <u>value</u> of the partial derivative $\partial f(x)/\partial x^i$ of f with respect to x^i at x can be obtained from (3.43) by setting

(3.44) $dx^i = 1$, $dx^1 = dx^2 = \ldots = dx^{i-1} = dx^{i+1} = \ldots = dx^\nu = 0$.

The total differential (3.43) thus contains all the information necessary to calculate each partial derivative $\partial f(x)/\partial x^i$, $i = 1,2,\ldots,\nu$. (Conversely, given this set of partial derivatives, one can form the differential df from (3.43).) As in the example given by (3.40)-(3.41), this gives a way to obtain a code list for $\partial f/\partial x^i$ from code lists for the function f and its differential df.

The expression (3.43) can be regarded as the product (in the matrix sense) of the <u>derivative</u> vector

$$(3.45) \qquad f'(x) = (\partial f(x)/\partial x^1 \quad \partial f(x)/\partial x^2 \quad ... \quad \partial f(x)/\partial x^\nu)$$

which is a $1 \times \nu$ matrix, or <u>row</u> <u>vector</u>, and the $\nu \times 1$ <u>column</u> <u>vector</u> of <u>differentials</u>

$$(3.46) \qquad dx = \begin{bmatrix} dx^1 \\ dx^2 \\ ... \\ dx^\nu \end{bmatrix} = (dx^1 \quad dx^2 \quad ... \quad dx^\nu)^T ,$$

where the superscript T denotes transposition. Thus, in matrix notation, (3.43) may be written simply as

$$(3.47) \qquad df = f'(x) \cdot dx ,$$

where $\cdot$ denotes matrix ("row-by column") multiplication. The differential df is also often seen expressed in vector notation in terms of the <u>gradient</u> <u>vector</u> $\nabla f(x)$, which is the transpose of the derivative vector,

$$(3.48) \qquad \nabla f(x) = f'(x)^T = \begin{bmatrix} \partial f/\partial x^1 \\ \partial f/\partial x^2 \\ ... \\ \partial f/\partial x^\nu \end{bmatrix} ,$$

a column vector. The <u>inner</u> (or <u>scalar</u>) product of column vectors $a, b \in R^\nu$ is denoted by (a,b), where

$$(3.49) \qquad (a,b) = \sum_{i=1}^{\nu} a^i * b^i = a^1 * b^1 + a^2 * b^2 + ... + a^\nu * b^\nu = a^T \cdot b ,$$

and * denotes ordinary multiplication. In this notation,

$$(3.50) \qquad df = (\nabla f(x), dx) = \nabla f(x)^T \cdot dx .$$

Formulas (3.47) and (3.50) are seen at once to be generalizations of the definition

$$(3.51) \qquad df = (\frac{df}{dx}) * dx$$

of the differential in ordinary single-variable calculus, achieved by replacing the ordinary derivative df/dx by the derivative vector (3.45) or the gradient vector

(3.48), and the operation of ordinary multiplication * by the matrix product · or the vector inner product (,), respectively. These formulas also generalize (3.36) to functions of more than two variables.

In the terminology of functional analysis, the derivative f'(x) of f at x, if it exists, is a _linear functional_ on R^ν; that is, f'(x) will map an _arbitrary_ vector dx ∈ R^ν into the _number_ df ∈ R. Results on limits in ordinary calculus associated with the definition of the derivative sometimes give the impression that the differentials df, dx should be "small", which is not necessarily the case here. All that is at stake here is a _linear transformation_ of dx into df, and dx can be taken to be arbitrary if x represents a set of ν independent variables. However, a word of warning is in order here about the concept of differentiability.

Warning! The existence of the first partial derivatives $\partial f/\partial x^i$ of f at x for i = 1,2,...,ν does _not_ imply that f is differentiable at x for ν > 1 unless a suitable limit condition is satisfied which will guarantee that the difference f(x + dx) - f(x) will be approximated "arbitrarily closely" by the differential df as the vector dx goes to the _zero_ _vector_ 0 = (0,0,...,0). To be more precise, let the _length_ of a vector h ∈ R^ν be defined by a suitable _norm_ ‖h‖ (see [62], [71] for more details; for the present, either the _maximum_ _norm_ $\|h\| = \max_{(i)} |h^i|$ or the usual Euclidean _norm_ $h = (h,h)^{1/2} = ((h^1)^2 + (h^2)^2 + ... + (h^\nu)^2)^{1/2}$ will be sufficient to consider). By an "arbitrarily close" approximation of the difference Δf = f(x + dx) - f(x) by the differential df as dx goes to zero, one possible definition is

(3.52) $$\lim_{\|dx\| \to 0} |f(x + dx) - f(x) - f'(x) \cdot dx| / \|dx\| = 0.$$

If f'(x) satisfies (3.52), then it is called the _Fréchet_ _derivative_ of f at x [71]. A less restrictive condition is obtained by setting dx = τh = $(\tau h^1 \ \tau h^2 \ ... \ \tau h^\nu)^T$ for h a _unit_ _vector_ (‖h‖ = 1) and requiring that for all unit vectors h,

(3.53) $$\lim_{\tau \to 0} \left| \frac{f(x + \tau h) - f(x)}{\tau} - f'(x) \cdot h \right| = 0 .$$

This condition characterizes the _Gâteaux_ _derivative_ of f at x [62]. Both of the above notions generalize the concept of the derivative of a function of a single variable to functions of several variables, and both are useful in numerical analysis. Although the main thrust of the discussion here is to develop techniques for the automatic computation of the partial derivatives $\partial f/\partial x^i$, some of the applications to be considered later will require that f is differentiable at least in the sense of (3.53). You have been warned.

Another convenience furnished by the use of differentials in order to obtain code lists for partial derivatives is that this method also allows the computation of what were called _semi-total_ derivatives above (see (3.4)) in a straightforward way. For example, suppose that

$$y = y(x,z),$$

(3.54)

$$f = f(x,y,z) .$$

From these formulas, one obtains the differentials

$$dy = \frac{\partial y}{\partial x} dx + \frac{\partial y}{\partial z} dz ,$$

(3.55)

$$df = \frac{\partial f}{\partial x} dx + \frac{\partial f}{\partial y} dy + \frac{\partial f}{\partial z} dz .$$

Substitution of the expression obtained for dy into the formula for df gives

(3.56)
$$df = (\frac{\partial f}{\partial x} + \frac{\partial f}{\partial y} \frac{\partial y}{\partial x}) dx + (\frac{\partial f}{\partial z} + \frac{\partial f}{\partial y} \frac{\partial y}{\partial z}) dz .$$

In (3.56), the coefficient of dx is the semi-total derivative $Df/\partial x$, and the coefficient of dz is $Df/\partial z$. Hence, by setting dx = 1, dz = 0 in (3.56), one gets df = $Df/\partial x$ as defined by (3.4). This technique may be applied to (3.6), which leads to the expressions

(3.57)
```
DU = DY + X*DZ + DX*Z
DV = (X*DU + U*DX - X*U*DY/Y)/Y
DG = 2*X*DX + EXP(V)*DV + DU ,
```

which are the differentials of the three lines of code (3.6) written for the evaluation of the function g(x,y,z) defined by the formula (3.5). Evaluation of the list (3.57) with the current values of X, Y, Z, and DX = 1, DY = DZ = 0 gives DG = DG/∂X = $\partial g(X,Y,Z)/\partial x$, the true value of the partial derivative of the function g with respect to x at the point (X,Y,Z). Of course, replacing the lines in (3.57) by their corresponding code lists would give the same result, because of the equivalence of formulas and code lists for the representation of functions. The important point is that whether done by derivatives or differentials, software for automatic differentiation must have the capability of producing correct derivatives of functions defined by several lines of code, at the option of the user.

The key features of differentiation by derivatives or differentials are that the output of the differentiator is a code list (simplified or not) of the same form as the input code list for the evaluation of the function to be differentiated, and the output code list (i) represents the desired derivative or differential correctly; (ii) can be executed interpretively or compiled into object code by the same interpreter or compiler which will accept the code list for the original function; and thus (iii) the output code list for the derivative may itself be differentiated with respect to any variable it contains by the differentiator, so that mixed partial derivatives of arbitrary order of the original function may be obtained by successive differentiations. The principal distinction between the use of differentials as compared to derivatives is that the code list obtained in differential form contains all the information necessary to compute derivatives of the function considered with respect to each of its variables, while the derivative method results in only the derivative of the function with respect to a single given variable each time it is applied.

To illustrate the output of a differentiator, suppose that the code list (3.40) has been simplified to obtain the following code list for DXF = f'_x(X,Y) by reference to Figure 3.2 and (3.41):

$$
\begin{aligned}
\text{T5} &= \text{Y**2} \\
\text{T6} &= 3*\text{T5} \\
\text{T7} &= \text{T6} + 6 \\
\text{DXT2} &= \text{COS}(X) \\
\text{DXT4} &= Y + \text{DXT2} \\
\text{DXF} &= \text{T7*DXT4} \; .
\end{aligned}
$$

(3.58)

In addition to simplification of the code list for DXF, the differentiator should re-label the lines in the simplified list (3.58) as, say, U1 = T5, U2 = T6, U3 = T7, U4 = DXT2, U5 = DXT4, DXF, so that the final output will be a list of the form

$$
\begin{aligned}
\text{U1} &= \text{Y**2} \\
\text{U2} &= 3*\text{U1} \\
\text{U3} &= \text{U2} + 6 \\
\text{U4} &= \text{COS}(X) \\
\text{U5} &= Y + \text{U4} \\
\text{DXF} &= \text{U3*U5} \; ,
\end{aligned}
$$

(3.59)

which is ready for execution, compilation, or further differentiation.

Another capability provided to the user of a differentiating compiler (that is, a compiler which includes formula differentiation as well as formula translation) is implicit differentiation. In many applications, a dependent variable y is not defined as a function of independent variables u, v, w, x, for example, by an explicit formula

(3.60) $y = y(u,v,w,x)$

involving library functions and arithmetic operations, but rather by an implicit relationship given by a formula such as

(3.61) $g(u,v,w,x,y) = 0$.

In theory, one can think of solving the equation (3.61) for y to obtain (3.60), however, this may be impractical in many cases, and there may even fail to be a formula which expresses y as a function of u, v, w, x, in terms of the given set of library functions and arithmetic operations. On the other hand, in terms of differentials of the implicit relationship (3.61), one has

(3.62) $\dfrac{\partial g}{\partial u}\, du + \dfrac{\partial g}{\partial v}\, dv + \dfrac{\partial g}{\partial w}\, dw + \dfrac{\partial g}{\partial x}\, dx + \dfrac{\partial g}{\partial y}\, dy = 0$.

Thus,

$$(3.63) \qquad dy = - (\frac{\partial g}{\partial u}/\frac{\partial g}{\partial y}) \; du - (\frac{\partial g}{\partial v}/\frac{\partial g}{\partial y}) \; dv - (\frac{\partial g}{\partial w}/\frac{\partial g}{\partial y}) \; dw - (\frac{\partial g}{\partial x}/\frac{\partial g}{\partial y}) \; dx \; ,$$

from which can be obtained not only dy but also, for example, $\partial y/\partial x$ by setting the differential vector (du, dv, dw, dx) = (0, 0, 0, 1), and so on. Since equation (3.62) is <u>linear</u> in the differentials du, dv, dw, dx, dy, it follows that one does not have to solve the generally <u>nonlinear</u> equation (3.61) for y and then differentiate the resulting solution (3.60) to obtain dy. In order to do implicit differentiation using an automatic differentiator, suppose, for example,

$$(3.64) \qquad\qquad\qquad G = g(U,V,W,X,Y) \; .$$

One gets from (3.43) that the code list for DG will represent a function which may be expressed as

$$(3.65) \qquad DG = DUG*DU + DVG*DV + DWG*DW + DXG*DX + DYG*DY$$

$$= (DUG \; DVG \; DWG \; DXG \; DYG) \cdot (DU \; DV \; DW \; DX \; DY)^T$$

$$= DG(DU,DV,DW,DX,DY) \; .$$

Thus,

$$(3.66) \qquad\qquad \frac{\partial g}{\partial y} = DYG = DG(0, 0, 0, 0, 1)$$

is the denominator required in (3.63), and the numerators may be obtained in the same way. If only a certain partial derivative of y is required, say $\partial y/\partial x$, then

$$(3.67) \qquad \frac{\partial y}{\partial x} = DXY = - DXG/DYG = - DG(0,0,0,1,0)/DG(0,0,0,0,1) \; .$$

Given simplified code lists for DXG and DYG, a code list for DXY can be obtained by the addition of the following two lines,

$$(3.68) \qquad\qquad\qquad T = (-1)*DXG$$
$$DXY = T/DYG \; .$$

(The multiplication by −1 in (3.68) could be considered to be performed by a library function called perhaps CHS; then, T = CHS(DXG) could replace the first line in the list (3.68).) Thus, this formulation of implicit differentiation requires only two lines following the code lists for DXG and DYG. Of course, if the code lists for DXG, DYG contain common entries, then some simplification can be done before the lines (3.68), or their equivalents, are adjoined to the list.

It has been shown that derivatives can be obtained from code lists for differentials by picking specific values for the differential vector (1 for the component corresponding to the variable of interest, 0 otherwise). It is also possible to obtain differentials from derivatives, so the two methods are equivalent in this sense. To do this, one may introduce a fictitious variable ∅, and consider all other variables to be functions of ∅. Then, for example, dx = x'(∅)*d∅, and so on. The partial derivatives of $f = f(x,y,z)$, for example, may then be obtained by setting $x = \emptyset$, $y'(\emptyset) = z'(\emptyset) = 0$, and similarly for the other variables in turn.

    3. <u>Nomenclature for code lists</u>. In Table III.1 below, formulas and code lists are given in differential form for the arithmetic operations and a few library functions. In order to give a precise description of how this table (which will sometimes be referred to as a <u>dictionary</u>) is used, it will be helpful to give a more formal characterization of the structure of a code list than the intuitive notion presented up to this point. Suppose that f_1, f_2, ..., f_n, f are obtained from a basic sequence for f by deleting data (variables, parameters, and constants), and then renumbering if necessary. The corresponding <u>code list</u> will consist of n + 1 <u>lines</u>, where each line is of the form

(3.69)   label = entry .

(The functions f_i, i = 1,2,...,n, in the basic sequence for f which remain after the deletion of data are, of course, restricted to the set of unary library functions and binary arithmetic operations permitted.)

    In a <u>simple</u> code list, the first n labels number the lines in the list in consecutive order in association with some symbol which identifies the list. For example, if <name> denotes the name of the function to be evaluated (such as F), then the ith label could have the form name <i> (or, as in some software, <i><name>), i = 1,2,...,n. Thus, if F is the function being represented, then the labels in a simple code list for F could have the form F1, F2, ..., or F001, F002, ..., or perhaps 001F, 002F, ... The last label in a simple code list is <name> to identify the function represented by the code list. In the example cited, <name> = F. The ith line in a simple code list will then have the label denoted by label(i), formed according to the above rules, i = 1,2,...,n,n+1. The corresponding entry, entry(i), can have one of the following two forms:

                              entry(i) = <LIB(arg(i))>

(3.70)   or

                              entry(i) = <left(i)><OP><right(i)> ,

where LIB denotes a library function, OP one of the arithmetic operations +, -, *, /, **, and the arguments arg(i), left(i), right(i) are restricted to be either constants, parameters, variables, or label(j) for some j < i, that is, the label of a previous line in the code list. An example of a simple code list is, of course, given by (2.3). Thus, an entry in a code list will not be a single piece of data.

    A <u>compound</u> code list is a sequence of simple code lists. A code list (simple or compound) which represents a function will be called a <u>function</u> code list. Each code list represents the function named by the label of its last line.

    Table III.1 gives simple code lists for the differentials of entries which involve the arithmetic operations or a few standard library functions. Insertion of these lists with suitable labels in place of the corresponding line in the code list for the function to be differentiated will give a compound code list which will be called a _differential_ code list, in which the differentials of the variables in the original list are regarded as new variables. By giving these differentials their proper values, a code list for the desired derivative can be obtained from the differential code list, as indicated earlier. This list, naturally, will be called a _derivative_ code list. For example, by setting DX = 1, DY = 0 in (3.40) and using the labeling system of Table III.1, one obtains the list (which is not a code list),

$$
\begin{aligned}
\text{DXT11} &= \text{X*0} \\
\text{DXT12} &= \text{Y*1} \\
\text{DXT1} &= \text{DXT11} + \text{DXT12} \\
\text{DXT21} &= \text{COS(X)} \\
\text{DXT2} &= \text{DXT21*1} \\
\text{DXT3} &= \text{DXT1} + \text{DXT2} \\
\text{DXT4} &= \text{DXT3} \\
\text{DXT51} &= \text{Y**1} \\
\text{DXT52} &= \text{2*DXT51} \\
\text{DXT5} &= \text{DXT52*0} \\
\text{DXT6} &= \text{3*DXT5} \\
\text{DXT7} &= \text{DXT6} \\
\text{DXF1} &= \text{T4*DXT7} \\
\text{DXF2} &= \text{T7*DXT4} \\
\text{DXF} &= \text{DXF1} + \text{DXF2} \ ,
\end{aligned}
$$

(3.71)

from which the derivative code list can be obtained. A method for doing this will be described below.

Of course, the list (3.71) could be obtained directly from the code list (2.3) by use of Table III.1 without forming the differential code list (3.40). If derivatives of a function with respect to several of its variables are required, then its function code list (for example, (2.3)) can be processed several times in this way to obtain lists of the form (3.71). However, as the differential code list contains all the information required to form derivative code lists with respect to each variable in the function code list, its formation as an intermediate step may be preferable to repeated processing of the function code list if a number (or all) of the partial derivatives of the function are desired. The production of a differential code list as an intermediate step in differentiation could then be provided as an option to the user.

The list (3.71) contains a number of unnecessary lines, which must be eliminated to obtain a code list for the derivative of F with respect to X. One approach to the elimination of such lines is to write the differentiating software in such a way that it will not form a line unless the conditions (3.70) are satisfied. This is essentially the method used in the programs CODEX and SUPER-CODEX, which will be described in Chapter 5. An alternative way to obtain the derivative code list is to go ahead with the formation of the list (3.71), and then eliminate unnecessary lines. The process of removing these lines is called packing, and the resulting list, such as (3.41) in this case, is called a packed code list. The labels in a packed code list are still numbered in ascending order, but not necessarily consecutively as in a simple code list. For example, (3.41) does not contain a line labeled DXT3 between DXT2 and DXT4.

Before the derivative DXF can be evaluated, the derivative code list for DXF must be preceded by the code list for the function F. The resulting compound code list will be called a complete code list for the derivative DXF. As in the example given above, this list may also contain superfluous lines, which may be eliminated by packing the list. Thus, (3.58) is a packed complete derivative code list, which can stand alone as a function code list. Given such a complete code list, the lines may be relabeled to obtain a simple code list, in this case (3.59), which will also sometimes be called an output code list.

The process of automatic differentiation described in this chapter may therefore include all or some of the following code lists:

         1°. The function code list;

         2°. The differential code list;

         3°. The derivative code list;

         4°. The complete derivative code list;

         5°. The packed complete derivative code list;

         6°. The output code list.

The distinction between the lists 2° and 3° depends essentially on the number of variables indicated by the user, and if only one variable is indicated, the list 3° will be produced if the differential of the variable is given the literal (constant) value one. Otherwise, a differential code list 2° could be generated as an intermediate step in obtaining (partial) derivatives. In the latter case, formation of the complete derivative code list requires preceding the differential code list with code lists defining all arguments and all differentials appearing in the differential code list. Because of the tendency of differentiation to produce long code lists, packing is a desirable feature of an automatic differentiator, and should be implemented as skillfully as possible in order to produce lists of minimal length. The choice between a function list 5° and a simple code list 6° as the final result is a matter of taste and the use to which the final list will be put, and can be left to the user.

A brief dictionary for the formation of differential code lists from function code lists is given in the following Table III.1. Conventions used in this table are:

(3.72)   <label> = T;   <arg> = U;   <left> = V;   <right> = W,

based on the notation (3.69), (3.70). For the differential of a constant,

(3.73)   D<constant> = 0,

where <constant> is the name of a literal constant, or a symbol not declared to be a variable (in other words, a parameter).

TABLE III.1.   A BRIEF DICTIONARY FOR THE FORMATION OF
DIFFERENTIAL CODE LISTS

Entry	Simple Code List for Differential of Label

A. Arithmetic Operations

(3.74)   T = V ± W

$$DT = DV \pm DW;$$

(3.75)   T = V*W

$$DT1 = V*DW$$
$$DT2 = W*DV$$
$$DT = DT1 + DT2;$$

(3.76)   T = V/W

$$DT1 = V*DW$$
$$DT2 = DT1/W$$
$$DT3 = DV - DT2$$
$$DT = DT3/W;$$

(3.77)   T = V**W

$$DT1 = LOG(V)$$
$$DT2 = DT1*DW$$
$$DT3 = W*DV$$
$$DT4 = DT3/V$$
$$DT5 = DT2 + DT4$$
$$DT = T*DT5.$$

B. Some Library Functions

(3.78)   T = EXP(U)

$$DT = T*DU;$$

(3.79)   T = LOG(U)

$$DT = DU/U;$$

(3.80)   T = SIN(U)

$$DT1 = COS(U)$$
$$DT = DT1*DU;$$

(3.81)   T = COS(U)

$$DT1 = SIN(U)$$
$$DT2 = -1*DT1$$
$$DT = DT2*DU;$$

(3.82)   T = ARCTAN(U)

$$DT1 = U**2$$
$$DT2 = 1 + DT1$$
$$DT = DU/DT2.$$

Naturally, the capabilities of an automatic differentiator are determined by

its dictionary of library functions, just as in the case of a formula translator.
Both should provide for the possibility for addition of new library functions as the
need arises for them. For example, for

$$C = \text{<constant>} \quad ,$$

it may be more convenient to consider $U**C$ to be a library function than to use the
general formula (3.77) for exponentiation with $V = U$, $W = C$. This could be done by
adding

(3.83) $T = U**C$ $DT1 = C - 1$

$DT2 = U**DT1$

$DT3 = C*DT2$

$DT = DT3*DU$

to the dictionary. The use of the formula (3.17) gives the more compact code list,

(3.84) $T = U**C$ $DT1 = C*T$

$DT2 = DT1/U$

$DT = DT2*DU$,

which is equivalent to the list obtained from (3.77) by setting $DW = 0$ and packing.
(The form (3.83) with $C = 2$ and packing was used in the example (3.40) and the following lists for the differential of $T = Y**2$, see also (3.95).) The lists (3.77)
and (3.84) may not give the correct value if $U = 0$, or $V = 0$ in (3.77), even
though the value of the exponential and its derivative are well defined. One way
to provide for possibilities such as this will be discussed later, in connection
with Taylor coefficients of exponential functions, and the differentiation of piecewise defined functions.

The code list 1° for the function being differentiated will also be called an
input code list. If the input list is a simple code list, then the differential
code list will ordinarily be a compound code list, since the dictionary gives simple
code lists consisting of several lines for the differentials of single lines in the
original function code list. Suppose that label = <name><i> for the line being processed in the input code list. Then, according to the convention adopted above, the
last line in the simple code list for the differential of the entry in the line being processed will be labeled D<name><i>, following intermediate lines with consecutive labels D<name><i>1, D<name><i>2, ..., D<name><i>k, where k depends on the particular entry. For example, in Table III.1, k = 3 for V/W, and k = 5 for V**W. The
labels of the intermediate lines are only referred to in the simple code list for
D<name><i>, and not elsewhere in the differential code list. It follows that as the
result of packing one of the simple sublists obtained from the dictionary to produce
the derivative code list, the result should be labeled D<name><i>, rather than by
the label of an intermediate line. For example, (3.75) with $DW = 1$, $DV = 0$ gives

(3.85) $DT1 = V*1$

$$DT2 = 0*W$$
$$DT = DT1 + DT2 .$$

The result of packing this sublist is

(3.86)
$$DT = V ,$$

and all references to DT in the following portion of the derivative code list should be replaced by references to V. Starting from the semi-packed list

(3.87)
$$DT1 = V$$
$$DT = DT1 ,$$

the second line can be omitted if all references to DT are replaced by references to DT1. Then, one obtains the (correct) result

(3.88)
$$DT1 = V ,$$

but then one has to go over the following code list again and replace all references to DT1 by references to V. The method of computing the literal value of a line, if possible, followed by substitution of that value for later references to the label of the line and deletion of the line in question is called _forward_ packing. Applying this to (3.85), the literal value DT1 = V is calculated by the packer. (A line is said to have a _literal value_ if its entry is a constant, parameter, variable, or the label of a previous line.) Forward substitution of the result (3.88) gives the partially packed code list

(3.89)
$$DT2 = 0*W$$
$$DT = V + DT2 ,$$

and computation of the literal value DT2 = 0 and forward substitution gives

(3.90)
$$DT = V + 0 ,$$

the literal value of which is (3.86). Note that the range of forward substitution of literal values for intermediate code lines DT1, DT2, ..., DTk is limited by the length of the simple code list in the dictionary for the differential of the entry. However, if a literal value is obtained for the final line in the dictionary code list (that is, for DT), then the entire remainder of the differential code list has to be scanned for references to DT, which have to be replaced by the literal value of DT before the line with label DT can be omitted from the code list. Forward packing of (3.71) gives

(3.91)
$$DXT2 = COS(X)$$
$$DXT3 = Y + DXT2$$
$$DXF = T7*DXT4 ,$$

where the variable of differentiation X has been inserted into the labels in the differential code list (3.40) to obtain the corresponding labels for the derivative code lists (3.71) and (3.91). To make the packed derivative code list (3.91) into

a complete derivative code list (which can be executed to evaluate the derivative),
it must be preceded by the code list for the function F, since an entry in (3.91)
refers to a label in the input function code list. Thus, going back to (2.3),

$$
\begin{aligned}
T1 &= X*Y \\
T2 &= SIN(X) \\
T3 &= T1 + T2 \\
T4 &= T3 + 4 \\
T5 &= Y**2 \\
T6 &= 3*T5 \\
T7 &= T6 + 6 \\
F &= T4*T7 \\
DXT2 &= COS(X) \\
DXT4 &= Y + DXT2 \\
DXF &= T7*DXT4 \;.
\end{aligned}
$$

(3.92)

This list can be packed by a different method, called <u>backward</u> packing. (It is
assumed that the input code list has already been packed.) The lines of the derivative code list are examined in turn, starting from the first line, for references
to the input function code list. Each line in the function code list found in this
way is checked for references to previous lines in the function code list, and so on.
Lines which do not have labels which are referred to directly or indirectly by entries in the derivative code list may then be omitted from the function code list
to obtain the packed complete derivative code list. Actually, the order in which
the lines in the derivative code list are examined is immaterial, so that this process could just as well start with the last line of the derivative code list and
work upward.

To illustrate backward packing, the lines labeled DXT2 and DXT4 in (3.92) do
not refer to the function code list. The line labeled DXF, however, refers to T7,
the line labeled T7 to T6, the line labeled T6 to T5, while the entry in the line
labeled T5 refers to Y and the literal constant 2. Thus, all lines except T5, T6,
and T7 can be omitted from the input function code list, giving finally

$$
\begin{aligned}
T5 &= Y**2 \\
T6 &= 3*T5 \\
T7 &= T6 + 6 \\
DXT2 &= COS(X) \\
DXT4 &= Y + DXT2 \\
DXF &= T7*DXT4
\end{aligned}
$$

(3.93)

as the packed complete derivative code list for DXF. This is a function code list
which will evaluate $DXF = \partial f(x,y)/\partial x$ for the given values $x = X$, $y = Y$ of the input
variables. It may be transformed into a simple code list by relabeling its lines
in consecutive order:

$$\begin{aligned}
U1 &= Y**2 \\
U2 &= 3*U1 \\
U3 &= U2 + 6 \\
U4 &= COS(X) \\
U5 &= Y + U4 \\
DXF &= U3*U5 \ .
\end{aligned}$$

(3.94)

This is an _output_ code list for DXF in a form suitable for listing or, of course, further processing by an interpreter, compiler, or differentiator.

In the code list (3.40) and the subsequent lists used as examples above, Y**2 was assumed to have been computed by a library function, and the corresponding entry added to the dictionary in Table III.1:

(3.95) $T = U**2$ $DT1 = 2*U$

                                       $DT = DT1*DU$.

In many cases, as in connection with the use of interval arithmetic [53], [55], addition of subroutines and differentials of simple functions such as this to the set of library functions and their differentials is usually worthwhile, even though the dictionary is lengthened.

4. Projects for Chapter 3. The projects listed below are suggestions based on the concepts presented above. They may be carried out in detail ranging from homework exercises to the development of software with commericial possibilities.

1°. Write a _coder_ for your favorite language (FORTRAN, PASCAL, etc.) to produce a simple function code list from a formula for the function.

2°. Write a program which will produce a differential code list as output, given a function code list and a list of variables as input.

3°. Write a program which will produce a packed derivative code list as output, given a differential code list and the name of the variable of interest as input.

4°. Write a program which will produce a packed complete derivative code list as output, given the function code list and a packed derivative code list.

5°. Write a program which will produce a function code list for the derivative of a function defined by a formula with respect to a given variable.

CHAPTER IV

GENERATION OF TAYLOR COEFFICIENTS

The automatic generation of Taylor coefficients of a function is a process related closely to differentiation. If f is a function of a single variable, then its
nth Taylor coefficient at x is defined to be

$$(4.1) \qquad f_n(x) = \frac{1}{n!} f^{(n)}(x), \quad n = 1, 2, \ldots,$$

where $f^{(n)}(x)$, as usual, denotes the nth derivative of f at x. Thus, the Taylor coefficients of a function are simply constant multiples of its derivatives. It is
also convenient to extend the definition (4.1) to n = 0 by defining

$$(4.2) \qquad f_0(x) = f(x),$$

which is consistent with (4.1) if the usual conventions $0! = 1$ and $f^{(0)}(x) = f(x)$
are adopted. Given the Taylor coefficients (4.1)-(4.2) of a function f, a familiar
chapter of mathematical analysis deals with its representation

$$(4.3) \qquad f(x) = \sum_{n=0}^{\infty} f_n(x_0)(x - x_0)^n$$

by a power series expansion about $x = x_0$, that is by its Taylor series. The utility
of this representation is based on the fact that if a positive number ρ exists such
that the infinite series (4.3) converges for $|x - x_0| < \rho$, then, within the circle
of convergence defined in this way, the results of a number of types of operations
applied to f, including differentiation and integration, may be obtained by computing with the terms of the Taylor series (4.3) to get the answer also in the form of
a convergent power series. One well-known and important application of Taylor series
is to the solution of initial-value problems (sometimes called Cauchy problems) for
ordinary differential equations [51], [53], [55], see also [10], [11].

Usually, actual computations are made with only a finite number of terms of the
Taylor series (4.3). One writes

$$(4.4) \qquad f(x) = \sum_{n=0}^{k} f_n(x_0)(x - x_0)^n + R_k(x; x_0) ,$$

where

$$(4.5) \qquad P_k f(x) = \sum_{n=0}^{k} f_n(x_0)(x - x_0)^n$$

is the Taylor polynomial of degree k obtained from f at x_0, and the remainder term
$R_k(x; x_0)$ may be written

(4.6)
$$R_k(x;x_0) = f_{k+1}(\theta x + (1 - \theta)x_0)(x - x_0)^{k+1} ,$$

where θ is some number satisfying $0 < \theta < 1$, or as

(4.7)
$$R_k(x;x_0) = (k+1)\int_0^1 f_{k+1}(\theta x + (1 - \theta)x_0)(1 - \theta)^k(x - x_0)^{k+1}d\theta .$$

The first of these formulas is sometimes called the <u>Lagrange</u> form of the remainder term, while (4.7) is referred to as the <u>Cauchy</u> form.

Calculations with the Taylor polynomial (4.5) require evaluation of the Taylor coefficients f_0, f_1, ..., f_k of f at the point $x = x_0$. The error made in the approximation of $f(x)$ by the Taylor polynomial $P_k f(x)$ can be estimated on the basis of (4.6) or (4.7) if the Taylor coefficient $f_{k+1}(z)$ can be obtained as a function of z, where

(4.8)
$$z = \theta x + (1 - \theta)x_0, \quad 0 \le \theta \le 1 ,$$

that is, z takes on values in the interval with endpoints x_0 and x. In some applications in numerical analysis, the Taylor coefficient f_{k+1} is generated automatically, after which it is evaluated by <u>interval arithmetic</u> [53], [55] to obtain upper and lower bounds for the remainder term. This combination of software for automatic generation of Taylor coefficients (or derivatives) and interval arithmetic gives the user the capability of obtaining error estimates in many problems as a direct result of the computation, and without the need to go through tedious analysis by hand (see [5], [27], [28], [29], [51], [52], [53], [55], [74] for elaboration and applications of these ideas).

Methods for the automatic generation of Taylor coefficients build directly on the ideas already introduced in the discussion of differentiation, particularly the use of code lists and dictionaries such as Table III.1 to obtain code lists for derivatives from function code lists. This is not surprising, since (4.1) shows that derivatives and Taylor coefficients are interchangeable from a mathematical point of view. From a computational standpoint, however, this interchange between derivatives and Taylor coefficients may give rise to some difficulties in practice. Given the derivatives of a function, its Taylor coefficients may be obtained easily and accurately from (4.1). On the other hand, the computation of derivatives from Taylor coefficients requires multiplication by n!, that is,

(4.9)
$$f^{(n)}(x) = n! \cdot f_n(x), \quad n = 0,1,2,\ldots,$$

and since n! increases rapidly with n, the errors in the Taylor coefficients will be multiplied by larger and larger numbers as one calculates the higher derivatives of f. This should be taken into account in the accuracy with which the Taylor coefficients are computed, if a certain accuracy is desired in the value obtained for the corresponding derivative by the use of (4.9).

1. <u>Subroutine call lists</u>. The idea behind automatic generation of Taylor coefficients may be visualized in terms of the previous discussion of differentiation
by reference to the Kantorovich graph of the calculation of a function, for example,
Figure 2.1. In the graph, each edge was assumed to transmit one piece of information, namely a numerical value obtained from the originating node, to the receiving
node. Each receiving node above the final node has one or two inputs consisting of
the values of constants (including parameters), variables, or outputs of other nodes.
The input value or values are processed at the node, and the result obtained from
the indicated operation or library function is transmitted as the output value from
that node along the indicated edge or edges of the graph. Suppose now that one permits each edge in the graph to carry <u>two</u> pieces of information, namely the value obtained from the originating node, and the value of the differential of that quantity. That is, the value of X and DX would be sent from the node labeled X in the
graph, the values of T4 and DT4 from node T4, and so on. Along with this, the processing capability of each node would have to be increased to produce the output
value and its differential from the input value(s) and differential(s).

Of course, this situation can be depicted by a new Kantorovich graph in which
the number of original edges is doubled, and each node is replaced by a subgraph
which will do the required processing. A subgraph corresponding to a multiplication
node in the original graph is shown in Figure 4.1.

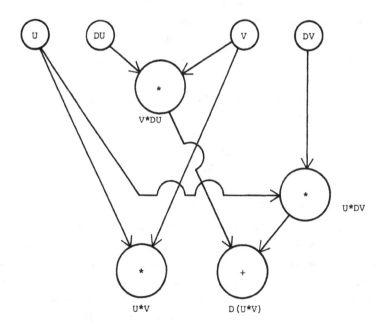

Figure 4.1. A Kantorovich Subgraph for the Value
and Differential of a Product.

Code lists for subgraphs of this type may be obtained directly from the dictionary by adding the line in the function code list directly before the simple code list for the differential of its entry. In the example depicted in Figure 4.1, the use of (3.75) from Table III.1 gives immediately

$$
\begin{aligned}
T &= U*V \\
DT1 &= U*DV \\
DT2 &= V*DU \\
DT &= DT1 + DT2 \ .
\end{aligned}
$$

(4.10)

If each line in a function code list, such as (2.3), is followed by the simple code list obtained from the dictionary for the differential of that line, then the result will be called a complete differential code list for the function F and its differential DF. This list may be generated as the code list for F is formed, rather than by processing the function code list later. The complete differential code list may also be processed by the techniques described in the previous chapter; in particular, code lists for partial derivatives of F may be obtained by assigning appropriate values to the differentials of the variables.

However, another approach to the differentiation of F may be taken, based on the fact that (4.10) can be regarded as a code list for a subroutine which computes the value of a product and its differential. This subroutine could be named

SUBROUTINE MULT(U,V,DU,DV,T,DT)

and similar names could be assigned to the other subroutines obtained from the code lists in the dictionary for other arithmetic operations and library functions. Thus, the code list (2.3) could be replaced by the subroutine call list

(4.11)

```
CALL MULT(X,Y,DX,DY,T1,DT1)
CALL SINE(X,DX,T2,DT2)
CALL ADDT(T1,T2,DT1,DT2,T3,DT3)
CALL ADDC(T3,4,DT3,T4,DT4)
CALL SQRE(Y,DY,T5,DT5)
CALL MULC(T5,3,DT5,T6,DT6)
CALL ADDC(T6,6,DT6,T7,DT7)
CALL MULT(T4,T7,DT4,DT7,F,DF) .
```

In the above list, it has been taken to be convenient to distinguish between binary addition (ADDT) and multiplication (MULT) of labels in general, and addition (ADDC) and multiplication (MULC) of literal constants as unary library functions, since the writing of the corresponding subroutines is simplified in the latter case. This refinement, while not necessary, is the kind of thing which should be considered in the construction of the subroutine library. A library function denoted by SQRE in (4.11) has also been introduced to compute $T = Y**2$ and $DT = 2*Y*DY$.

The notation in (4.11) has been chosen to resemble FORTRAN. If subroutines are called procedures, then (4.11), written in the corresponding vernacular, would be called a _procedure list_. The point is that if the code lists in the dictionary of differentials have been converted into subroutines or procedures, then (4.11) can be used instead of the complete differential code list to compute the value of the function $f = f(x,y)$, its differential df, either or both of the partial derivatives $\partial f/\partial x$ and $\partial f/\partial y$, or semitotal derivatives if dx or dy are defined in terms of other variables. The same observations apply to functions of more than two variables.

The method to be described now for the generation of Taylor coefficients follows the approach of construction of subroutine call lists rather that the technique of processing code lists. However, it should be noted that the _process_ of formation of the function code list (2.3) is followed in the formation of the subroutine call list (4.11), and, indeed, in the production of machine code for the evaluation of functions defined by formulas. Thus, the coder is still the key item of software in the differentiation process.

Another observation which may be made concerning the subroutine call list (4.11) is that _its_ _length_ _is_ _equal_ _to_ _the_ _length_ _of_ _the_ _function_ _code_ _list_ (2.3). If the total number of lines in the subroutine library for values and differentials is S and the code list for the function to be differentiated has L lines, then at most S + L lines of subroutine calls and library code will be required for the evaluation of derivatives (that is, first partial derivatives) of the function with respect to its variables. This shows that the coding for differentiation of functions may remain fairly compact, at least for first derivatives, if a suitable library of subroutines is available. A situation of common occurence in ordinary computing, of course, is the reduction of the bulk of a program by the judicious introduction of subroutines, so that this observation should come as no surprise.

The ability to compute higher derivatives in a compact form can also be based on the use of suitable subroutines. This also applies, of course, to the calculation of Taylor coefficients. In what follows, the construction of subroutines for generation of Taylor coefficients will be described on the basis of recursion formulas. In this instance, all variables, and hence the function being considered, will be assumed to be functions of a single (possibly fictitious) variable. This relates to one of the primary applications of Taylor series, which is the solution of systems of ordinary differential equations arising in celestial mechanics and other dynamical problems. Extensions of the methodology to problems involving several independent variables will be indicated later in Chapter 6.

2. _Recursion formulas for Taylor coefficients_. The idea of recursive generation of Taylor coefficients goes back a long ways, and is implicit in _Leibniz'_ _rule_ for calculating the derivatives of the product of two functions. Consider

$$(4.12) \qquad\qquad f(x) = g(x) * h(x) .$$

Successive differentiations of $f(x)$ give

$$f'(x) = g(x) * h'(x) + g'(x) * h(x),$$

$$(4.13) \qquad f''(x) = g(x)*h''(x) + 2*g'(x)*h'(x) + g''(x)*h(x) ,$$

$$f'''(x) = g(x)*h'''(x) + 3*g'(x)*h''(x) + 3*g''(x)*h'(x) + g'''(x)*h(x) ,$$

... ...

and so on. The general formula is easily recognized to be

$$(4.14) \qquad f^{(j)}(x) = \sum_{i=0}^{j} \binom{j}{i}*g^{(i)}(x)*h^{(j-i)}(x) ,$$

where $\binom{j}{i}$ denotes the __binomial coefficient__

$$(4.15) \qquad \binom{j}{i} = \frac{j!}{i!(j-i)!} , \quad i = 0,1,2,\ldots,j,$$

with $0! = 1$, as usual. In terms of Taylor coefficients instead of derivatives of g and h, (4.14) is simply

$$(4.16) \qquad f^{(j)}(x) = \sum_{i=0}^{j} j!*g_i(x)*h_{j-i}(x) ,$$

where (4.15) and the definition (4.1) of Taylor coefficients have been used. Division of both sides of (4.16) by j! and the use of (4.1) again yields finally

$$(4.17) \qquad f_j(x) = \sum_{i=0}^{j} g_i(x)*h_{j-i}(x) ,$$

which is a formula of appealing simplicity. Given the Taylor coefficients of g and h up to order k, the formula (4.17) may be used for j = 0,1,...,k to obtain the Taylor coefficients f_0, f_1, ..., f_k of f = g*h at the same point x. In Table IV.1 below, a number of formulas of the type (4.17) are given, corresponding to arithmetic operations and some selected library functions. Subroutines (or procedures) for performing the calculations indicated by these formulas may be prepared without difficulty.

Supposing that a library of subroutines to generate Taylor coefficients of functions including the arithmetic operations and certain specified library functions is available, the method of generating Taylor coefficients of a function defined by a code list can again be visualized by use of the corresponding Kantorovich graph. Returning to the example (2.3), now let each edge in the graph depicted in Figure 2.1 be strengthened so that it will be capable of transmitting a __vector__ of Taylor coefficients T = (T(0),T(1),...,T(K)) with its coefficients calculated at the __originating node__ T of the graph and, at the __destination node__, the incoming information received along the one or two incoming edges of the graph would be processed to give a vector of Taylor coefficients for further transmission or as the final result. For example, the node labeled T3 in Figure 2.1 would receive the vector

$$(4.18) \qquad T1 = (T1(0),T1(1),\ldots,T1(K)) ,$$

that is, the Taylor coefficients of T1 = X*Y from node T1, and the vector

(4.19)   T2 = (T2(0),T2(1),...,T2(K))

of Taylor coefficients of the function T2 = SIN(X) from node T2, and the output of
node T3 would be the vector

(4.20)   T3 = T1 + T2 =(T1(0) + T2(0),T1(1) + T2(1),...,T1(K) + T2(K))

of Taylor coefficients of the sum, using (4.22) in Table IV.1 below.

Instead of single values of the variables X, Y, ... appearing in the formula
for the function being processed, vectors of Taylor coefficients have to be supplied
to the program. For example, if Taylor coefficients are desired in terms of the
variable X (or the corresponding partial derivatives with respect to X), these starting vectors would have the form X = (X,1,0,...,0), Y = (Y,0,0,...,0), ... Similarly, if Y is the variable of interest, then X = (X,0,0,...,0), Y = (Y,1,0,...,0), ...
would be the corresponding starting vectors.

In using the above method for generation of Taylor coefficients, the code list
(2.3) would be replaced by a list of the form (4.11) of calls to library subroutines
or procedures. This list, as before, would be the same length as the original code
list for the function. However, an examination of the complexity of the process
should consider the amount of computation done in each subroutine in the library as
a function of the order K of the highest Taylor coefficient computed. (The list
(4.11) essentially corresponds to the case K = 1.) Looking, for example, at multiplication, formula (4.17) indicates that

(4.21) $M = \frac{1}{2}(K + 1)(K + 2)$

products are required in the computation of the Taylor coefficients of a product up
to and including order K. This implies a polynomial, rather than an exponential,
growth in the number of operations required. Other formulas for the recursive calculation of Taylor coefficients give similar results, see [55], §3.4.

TABLE IV.1. A SHORT TABLE OF RECURSION FORMULAS FOR TAYLOR COEFFICIENTS

	Function	Taylor Coefficients
A.	Arithmetic Operations	
(4.22)	$T = V \pm W$	$T(J) = V(J) \pm W(J);$
(4.23)	$T = V*W$	$T(J) = \sum_{I=0}^{J} V(I)*W(J-I);$
(4.24)	$T = V/W$	$T(J) = \{V(J) - \sum_{I=1}^{J} T(I-1)*W(J-I+1)\}/W(0);$
(4.25)	$T = V**W$	$T(J) = [EXP(W*LOG(V))](J);$

TABLE IV.1.   (CONTINUED)

Function	Taylor Coefficients

B. Some Library Functions

(4.26) $T = EXP(U)$

$T(0) = EXP(U(0))$,

for $J = 1,2,\ldots,K$,

$$T(J) = \sum_{I=0}^{J-1} ((J-I)/J)*T(I)*U(J-I);$$

(4.27) $T = LOG(U)$

$T(0) = LOG(U(0))$,

$T(1) = U(1)/U(0)$,

for $J = 2,3,\ldots,K$,

$$T(J) = \{U(J) - \sum_{I=0}^{J-1} ((J-I)/J)*U(J-I)*T(K) \}/U(0);$$

(4.28) $T = SIN(U)$

$\quad\quad\quad Z = COS(U)$

$T(0) = SIN(U(0))$,

$Z(0) = COS(U(0))$,

for $J = 1,2,\ldots,K$,

$$T(J) = \sum_{I=0}^{J-1} ((J-I)/J)*Z(I)*U(J-I),$$

$$Z(J) = - \sum_{I=0}^{J-1} ((J-I)/J)*T(I)*U(J-I);$$

(4.29) $T = ARCTAN(U)$

$\quad\quad\quad Z = 1/(1 + U**2)$

$T(0) = ARCTAN(U(0))$,

$Z(0) = 1/(1 + U(0)**2)$,

for $J = 1,2,\ldots,K$,

$$T(J) = \sum_{I=0}^{J-1} ((J-I)/J)*Z(I)*U(J-I).$$

In (4.25), $[EXP(W*LOG(V))](J)$ denotes the Jth Taylor coefficient of the function $V**W = EXP(W*LOG(V))$, which can be obtained by use of (4.26), (4.23), and then (4.26). Similarly, the Taylor coefficients $Z(J)$, $J = 1,2,\ldots,K$, required in (4.29) can be found from (4.23) with $V = 1$, $W = 1 + U**2$. As indicated previously, it is better to treat $U**2 = SQRE(U)$ as a library function, or use the method for constant exponents to be discussed in §3 below, rather than (4.25) for this simple function. An alternative, of course, is to set $U**2 = U*U$ and use (4.23); this is perfectly satisfactory in ordinary arithmetic, but is not suitable for interval arithmetic, as will be explained below.

Subroutines corresponding to the formulas in Table IV.1 can be coded without difficulty, and given appropriate names. For example,

(4.30)   SUBROUTINE ADDT(K,V,W,T)

could be called to perform the computations indicated by (4.22) in the table, with similar names and calling sequences for the other operations and functions cited. Given the code list for a function, a subroutine call list of the form (4.11) could then be made which would compute the value of the function at a point x, and a specified number K of its subsequent Taylor coefficients at that point.

As indicated in (4.11), there is an advantage to be gained by considering arithmetic operations in which one of the operands is a constant to be unary library functions, rather than use the general binary forms (4.22)-(4.25). This is particularly true in the case of generation of Taylor coefficients, since for $f(x) \equiv C$, a constant, one has

(4.31) $f_0(x) = C, \quad f_j(x) = 0, \quad j = 1,2,\ldots,$

independently of the value of x. The corresponding vector of Taylor coefficients is thus $C = (C,0,0,\ldots,0)$. It follows that many of the computations in the formulas (4.22)-(4.25) would be done with one of the operands equal to zero. This wasteful procedure may be avoided simply by programming the appropriate formulas into subroutines. As an example, in the case of a constant multiplier,

(4.32) $T = C*U \quad \text{or} \quad T = U*C$,

one can use the formula

(4.33) $T(J) = C*U(J)$

instead of (4.23). The resulting subroutine could be named

(4.34)   SUBROUTINE MULC(K,C,U,T) ,

where the general subroutine based on (4.23), which would also work in the case of a constant multiplier, but not as efficiently, could be called

(4.35)   SUBROUTINE MULT(K,V,W,T) .

Appropriate formulas for the generation of Taylor coefficients in the case that one argument is constant will be given below in Table IV.2 for +, -, *, /, and discussed in more detail for ** in §3.

As in (4.11), the list of subroutine calls for generation of the Taylor coefficients of a function up to some given order K will be exactly the same length as the code list for the function which has been produced by the coder. Since the subroutine library also consists of a fixed amount of code, this makes the coding of generation of Taylor coefficients fairly compact, and independent of the order desired. There is a problem of increase in storage requirements, however, although this is only linear with respect to the number of Taylor coefficients desired. Since each processing node in the Kantorovich graph, that is, each line in the code list for the function, generates a vector

(4.36) $$T = (T(0), T(1), \ldots, T(K))$$

of order $K + 1$, then, if there are N lines in the function code list, there is the possibility that $N*(K + 1)$ storage locations will be needed for the Taylor coefficients computed. The working storage needed by the subroutine library may also increase linearly with K, and, finally, the input variables will be given in the form of $K + 1$ dimensional vectors of their Taylor coefficients in terms of some basic variable $\emptyset$. Thus, including the Taylor coefficients of X and Y, the function defined by the eight-line code list (2.3) could require $10*(K + 1)$ storage locations for intermediate and final results. However, referring to the Kantorovich graph in Figure 2.1, it is seen that once T3 has been calculated, the values of T1 and T2 are no longer needed, and the corresponding storage locations could be reused, for example, by T4 and T5. The subroutine call list

(4.37)
```
CALL MULT(K,X,Y,T1)
CALL SINE(K,X,T2)
CALL ADDT(K,T1,T2,T3)
CALL ADDC(K,4,T3,T1)
CALL SQRE(K,Y,T2)
CALL MULC(K,3,T2,T3)
CALL ADDC(K,6,T3,T2)
CALL MULT(K,T1,T2,F)
```

requires only $6*(K + 1)$ storage locations, since the arrays T1, T2, T3 have been reused. This process of storage packing can be done on the basis of the code list for the function, in this case, (2.3). Starting with T1 and scanning the entire list, it is found that the last reference to T1 is in the line labeled T3. Hence, the next label, T4 in this case, can be changed to T1 in the rest of the list. Going on to T2, it is found that this label is also not needed after the line labeled T3. The next line after T3 which has not been relabeled is T5, so the label T2 is reassigned to it, and so on. This corresponds to rewriting the function code list (2.3) as

(4.38)
```
T1 = X*Y
T2 = SIN(X)
T3 = T1 + T2
T1 = T3 + 4
T2 = Y**2
T3 = 3*T2
T2 = T3 + 6
F = T1*T2 .
```

Actually, the results $F(J)$ could be stored in the locations reserved for $T3(J)$, thus reducing the overall storage requirements by a further $K + 1$ locations. The above procedure, however, requires great care in implementation, since one must keep track of how many times a line has been relabeled as well as references to each label in the remainder of the list.

A simpler approach to the reduction of storage may be based on the idea of indirect addressing. Some arrays A1, A2, ... of K + 1 storage locations are made available to the program as needed, and each used is assigned a current label from the corresponding line in the code list. As soon as a line is reached which follows all references to a given label, the corresponding storage locations are released, and become available to be assigned another current label and store the results of that line. The label of each successive line, then, is assigned to the first available array from the sequence A1, A2, ... Thus, when executing the subroutine call list corresponding to (2.3), as soon as the line labeled T4 is reached, the arrays A1, A2, which had current labels T1, T2, respectively, would be released, the current label T4 would be assigned to the array A1, and so on.

Of course, if any of the labels in the function code list are referenced outside that list, then the storage locations assigned to them cannot be released. This requires the programmer to distinguish between external and internal labels in a code list, and apply storage packing only to results which are strictly internal and not needed elsewhere. An example of such strictly internal results is furnished by the intermediate values produced by library subroutines. Coding formula (4.25) for the generation of Taylor coefficients of the exponentiation operation **, one has

(4.39)
$$
\begin{aligned}
U1 &= LOG(V) \\
U2 &= W*U1 \\
T &= EXP(U2) \ ,
\end{aligned}
$$

from which it follows that the Taylor coefficients $U1(J)$ are obtained from (4.27), $U2(J)$ from (4.23), and finally $T(J)$ from (4.26). After the numbers $U2(J)$ have been calculated, the values $U1(J)$ are no longer needed, so the final results $T(J)$ of this subroutine may be stored in the array assigned temporarily to $U1(J)$.

As indicated previously, the use of the general formulas in Part A of Table IV.1 for the generation of Taylor coefficients of functions resulting from arithmetic operations can involve wasteful calculations if one of the arguments is a constant, and this case is handled better by special subroutines. The formulas for the operations of addition, subtraction, multiplication, and division with one argument constant are given in Table IV.2 below.

TABLE IV.2. TAYLOR COEFFICIENTS FOR ADDITION, SUBTRACTION,
MULTIPLICATION, AND DIVISION WITH ONE CONSTANT ARGUMENT

Function	Taylor Coefficients
(4.39) $T = U + C = C + U$	$T(0) = C + U(0)$, for $J = 1,2,...,K$, $T(J) = U(J)$;

TABLE IV.2 (CONTINUED)

	Function	Taylor Coefficients
(4.40)	$T = U - C$	$T(0) = U(0) - C$, for $J = 1,2,\ldots,K$, $T(J) = U(J)$;
(4.41)	$T = C - U$	$T(0) = C - U(0)$, for $J = 1,2,\ldots,K$, $T(J) = - U(J)$;
(4.42)	$T = C*U = U*C$	$T(J) = C*U(J)$;
(4.43)	$T = U/C$	$T(J) = U(J)/C$;
(4.44)	$T = C/U$	$T(0) = C/U(0)$,

for $J = 1,2,\ldots,K$,

$$T(J) = - \{ \sum_{I=1}^{J} T(I-1)*U(J-I+1) \}/U(0).$$

3. <u>Exponentiation with one constant argument</u>. Use of the general operation of exponentiation **, coded as (4.39), for evaluation, differentiation, or generation of Taylor coefficients, can have some undesirable consequences in ordinary computation in the simple and frequently encountered cases that one of the arguments is a constant, that is, if

(4.45) $\qquad\qquad\qquad T = U**C$

or

(4.46) $\qquad\qquad\qquad T = C**U,$

where C denotes a constant. For example, the computation of X**4 cannot be carried out by (4.39) unless the input value of X is positive, as a zero or negative value of the argument should be rejected by the library subroutine for the logarithm. However, X**4 is certainly defined for all real numbers X, and is in fact analytic everywhere, so it is reasonable to expect it to be evaluated and differentiated correctly whenever it is encountered in a formula.

Since functions of the form $f(x) = x^c$ are among the first encountered in calculus, the impression is formed that they are extremely simple, and it may come as a surprise that they present special computational problems, and must be treated with care to get correct results. However, an examination of this class of functions in more detail reveals that their mathematical properties are strongly dependent on the nature of the constant c. There are, in fact, four cases: (i) If c is a positive integer, then for $c = n$, $f(x) = x^n$ is analytic everywhere and $f(0) = 0$; (ii) $c = 0$, in which case $f(x) = x^0 \equiv 1$ independently of the value of x; (iii) If $c = -n$ is a negative integer, then $f(x) = x^{-n}$ exists and is differentiable everywhere except at $x = 0$, and (iv) if c is a nonintegral real number, then $f(x) = x^c$ is not defined for negative x, unless $c = p/q$, where q is an odd integer, and $f(0)$ is also not defined unless c > 0, in which case one may take f(0) = 0. The behavior of this "simple" function is thus somewhat more complicated than may be assumed at first sight. This section will consequently be devoted to some results connected with the evaluation of the function (4.45), and also (4.46), including generation of the appropriate Taylor coefficients.

To take care of case (ii) first, if C = 0, then one assigns the value

$$(4.47) \qquad\qquad\qquad T = 1$$

to (4.45), and this value is then a constant in all subsequent evaluations or differentiations, and may be treated accordingly, including in processing by the packer. Taking case (i) next, suppose that C = N is a positive integer. For N = 1,

$$(4.48) \qquad\qquad\qquad T = X,$$

which is simply a reference to a variable, and can be handled in the same way as (4.47) by the packer, if desired. If N > 1, then a simple and straightforward way to evaluate

$$(4.49) \qquad\qquad\qquad T = X**N$$

is to use a subroutine which actually generates the lines of the code list

$$T1 = X*X$$
$$T2 = X*T1$$

(4.50)
$$\dots \quad \dots \quad \dots$$
$$T = X*T(N-2) \ ,$$

where it is understood, if necessary, that T0 = X. The list (4.50) requires N − 1 multiplications in order to calculate the value of (4.49). Furthermore, (4.50) may be differentiated line by line using (3.75) from Table III.1 to obtain DT, or (4.23) from Table IV.1 may be applied line by line if the Taylor coefficients T(J) of (4.49) are desired.

The same considerations apply almost verbatum to the more general expression

$$(4.51) \qquad\qquad\qquad T = U**N,$$

where, for N = 0,

$$(4.52) \qquad T = 1, \quad DT = 0, \quad T(0) = 1, \quad T(2) = T(3) = \dots = T(K) = 0,$$

and, for N = 1,

$$T = U, \quad DT = DU, \quad T(J) = U(J) \text{ for } J = 0,1,2,\dots,K.$$

The algorithm (4.50) could then be used, _mutatus mutandi_, to obtain T, DT, or the Taylor coefficients T(J), J = 0,1,2,...,K of the simple _power function_ (4.51) by using the appropriate formulas. However, as is well-known, there is a more efficient way to compute the Nth power of a function than (4.50). By the method of _repeated squaring_, the number of multiplications required can be reduced to the order of $\log_2 N$.

To implement this method, it is helpful to have a library subroutine for the function defined by $f(x) = x^2$; suppose, for example, that this subroutine is referred to as SQRE(X). For

(4.53)   T = SQRE(U),

one could write the code list

(4.54)   T = U*U,

and use the formulas for differentials or Taylor coefficients of products to produce the corresponding quantities for the function code list entry SQRE(U). This would be perfectly satisfactory if ordinary arithmetic is being used; however, if interval arithmetic is employed in the execution of the calculation [53], [55], then one has

(4.55) $[-1,1]*[-1,1] = [-1,1]$,

while the _interval_ _extension_ [53], [55] of $f(x) = x^2$ gives

(4.56) $[-1,1]^2 = [0,1]$,

since the set of squares of the real numbers belonging to a given interval is an interval which does not contain any negative numbers. Since the interval (4.56) is smaller than the interval (4.55) produced by interval multiplication, then this is the result that one would want the squaring subroutine SQRE to produce in case interval arithmetic is used.

Suppose now that the positive integer N, expressed in the _binary_ system, has ℓ digits (bits), where $\ell \leq \log_2 N + 1$. Then, the integer N is of the form

(4.57) $N = 2^\ell + \varepsilon_{\ell-1}*2^{\ell-1} + \ldots + \varepsilon_1*2 + \varepsilon_0$,

where each of the numbers ε_i, $i = 0, 1, \ldots, \ell-1$, is equal to either 0 or 1. The code list

(4.58)
$$
\begin{aligned}
T1 &= SQRE(U) \\
T2 &= SQRE(T1) \\
\ldots \quad & \ldots \quad \ldots \\
T\ell &= SQRE(T(\ell-1))
\end{aligned}
$$

or the corresponding subroutine using repeated calls to SQRE may be formed. If the result of packing the code list

(4.59)
$$
\begin{aligned}
S1 &= \varepsilon_0*U \\
S2 &= \varepsilon_1*T1 \\
\ldots \quad & \ldots \quad \ldots \\
S\ell &= \varepsilon_{\ell-1}*T(\ell-1)
\end{aligned}
$$

contains nonzero lines, say R1, R2, ..., Rr, then the code list

$$Q1 = R2*R1$$
$$Q2 = R3*Q1$$

(4.60)

$$\cdots \quad \cdots \quad \cdots$$
$$Q(r-1) = Rr*Q(r-2)$$
$$T = T\ell*Q(r-1) \ ,$$

which contains $r \leq \ell$ multiplications, will give the result (4.51). (If all the entries in the list (4.59) are zero, then, of course, $T = T\ell$.) The number of nonzero digits in the binary expression for N is called the __weight__ $\mu(N)$ of N in coding theory; one has $r = \mu(N) - 1$ as the number of multiplications required to compute the power function (4.51) after the ℓ squarings in (4.58).

The formulas for the differential and the Taylor coefficients of the library function SQRE are very simple. Thus, for

(4.61)
$$T = SQRE(U) \ ,$$

the differential code list is simply

(4.62)
$$DT1 = U*DU$$
$$DT = 2*DT1 \ .$$

(An alternative form of the second line in (4.62) is DT = DT1 + DT2.) Subroutines for the calculation of the Taylor coefficients of T may be obtained from the formulas

(4.63)
$$T(J) = 2*\{ \sum_{I=0}^{M-1} U(I)*U(J-I) \} + SQRE(U(M)) \text{ for } J = 2*M \text{ even,}$$

and

(4.64)
$$T(J) = 2*\{ \sum_{I=0}^{M} U(I)*U(J-I) \} \text{ for } J = 2*M + 1 \text{ odd,}$$

or, of course, simply from (4.23) with V = W = U. The above forms, however, require fewer multiplications, and (4.63) is preferable in connection with the use of interval arithmetic, as mentioned above. The coding of the exponentional function with positive integral exponent N by the use of repeated squaring appears in the discussions by Reiter [80] and Kedem [37], [95], of the automatic generation of Taylor coefficients.

For C = -N, a negative integer, the exponential

(4.65)
$$T = U**-N$$

can, of course, be coded as

(4.66)
$$T1 = U**N$$
$$T = 1/T1 \ ,$$

and the method of repeated squaring (4.58)-(4.60) can be used to calculate the first line. (In (4.60), it should be understood that Q0 = R1, that is, if (4.59) contains only one nonzero line of code, then $T = T\ell*R1$.) The Taylor coefficients $T1(J)$ of the first line of (4.66) are then obtained immediately from (4.63)-(4.64), or $DT1$ from (4.62) if differentials are desired. From (4.66),

(4.67)
$$
\begin{aligned}
DT11 &= SQRE(T) \\
DT12 &= DT1*DT11 \\
DT &= -1*DT12
\end{aligned}
$$

gives the differential DT in terms of the differential $DT1$ of the power function (4.51) with a positive integral exponent. Similarly, by a simple adaptation of (4.44), the Taylor coefficients $T(J)$ of (4.65) are given in terms of $T1(J)$ by

(4.68)
$$
T(0) = 1/T1(0),
$$
$$
\text{for } J = 1,2,\ldots,K,
$$
$$
T(J) = -\{ \sum_{I=1}^{J} T(I-1)*T1(J-I+1) \}/T1(0) .
$$

As an alternative to (4.66), equivalent results may be obtained by coding (4.65) as

(4.69)
$$
\begin{aligned}
T1 &= 1/U \\
T &= T1**N ,
\end{aligned}
$$

and use of the formulas and techniques already developed for positive integral powers. The formulation (4.66) is cited by Kedem [95]. Since $f(x) = x^{-n}$ is not analytic at $x = 0$, it is equally valid to use the formulas

(4.70)
$$
T(0) = U(0)**-N,
$$
$$
\text{for } J = 1,2,\ldots,K,
$$
$$
T(J) = \{ \sum_{I=0}^{J-1} ((-N*(J-I)-I)/J)*T(I)*U(J-I) \}/U(0),
$$

as given by Reiter [80]. The equivalent code list for the differential DT is given, of course, by (3.84).

Now, the above discussion covers (4.45) in case (iii), so that in combination with the earlier treatment of cases (i) and (ii), this takes care of integral values of C (positive, negative, or zero). The other possible situation, namely, that C is a nonintegral real number, can be dealt with by use of (3.84) and the equivalent of (4.70) to get

(4.71)
$$
T(0) = U(0)**C,
$$
$$
\text{for } J = 1,2,\ldots,K,
$$
$$
T(J) = \{ \sum_{I=0}^{J-1} ((C*(J-I)-I)/J)*T(I)*U(J-I) \}/U(0),
$$

which will give correct results except possibly at $U = U(0) = 0$, assuming that the value $U(0)**C$ can be obtained. To be more precise, (4.71) is always applicable if

$C < 1$, since the corresponding function is not differentiable at $x = 0$, in the sense that a limit of derivatives at positive real numbers exists as $x \to 0$. In case $C > 1$, however, it may be desired to obtain this limiting value for the differential, in which case (3.83) may be used, or to compute the corresponding results for the Taylor coefficients up to some order $K < C$, which cannot be done directly by formula (4.71). In order to extend the method given by Reiter [80] and Kedem [37], [95], to this case, <u>symbolic division</u> may be used. Define

(4.72)
$$T(0) = U(0)**C$$
$$S(0) = U(0)**(C-1) \ ,$$

so that $S(0)$ is actually the quotient $T(0)/U(0)$. Given $S(0),\ldots,S(J-1)$, a code list for $T(J)$ may be formed by coding the formula

(4.73)
$$T(J) = \sum_{I=0}^{J-1} ((C*(J-I)-I)/J)*U(J-I)*S(I) \ .$$

The code list for $T(J)$ will contain one or more lines in which the entries are of the form $U(0)**D$, where $D = C - M$, and M is an integer from the set $1,2,\ldots,J$. The code list for $S(J) = T(J)/U(0)$ is formed by replacing each entry $U(0)*D$ in the code list for $T(J)$ by $U(0)**(D-1)$, thus performing the division by $U(0)$ symbolically. This procedure may be continued until the desired value of $T(K)$ is obtained.

This method of symbolic division may also be used in case the exponent is an integer, as suggested by (3.83). From a practical point of view, the logarithm-antilogarithm (or logarithm-exponential) routine may be more convenient to code or faster in execution than repeated squaring, provided that suitable modifications can be made to make it an analytic function for positive integral exponents, and the same for negative integral exponents when the argument is outside a neighborhood of zero. Supposing that N is a positive integer, and

(4.74)
$$N \equiv P \pmod 2 \ ,$$

that is, $P = 0$ if N is even, and $P = 1$ if N is odd, then the exponential function (4.51) should be computed as

(4.75)
$$T = 0 \text{ if } U = 0,$$

otherwise,

(4.76)
$$
\begin{aligned}
T1 &= SIGN(U) \\
T2 &= T1*U \\
T3 &= LOG(T2) \\
T4 &= N*T3 \\
T5 &= EXP(T4) \\
T6 &= T1**P \\
T &= T6*T5 \ .
\end{aligned}
$$

The function $SIGN(U)$ referred to in (4.76) is $+1$ if $u > 0$, -1 if $u < 0$, 0 if $u=0$.

The last situation to be considered is (4.46), that is, exponentiation with a constant base C. Fortunately, this is handled simply by the logarithm-antilogarithm method (4.39). Here

$$
\begin{aligned}
T1 &= LOG(C) \\
T2 &= T1*U \\
T &= EXP(T2)
\end{aligned}
$$

(4.77)

Since T1 is a constant, the Taylor coefficients T(J) of T are given by

(4.78)

$$
\begin{aligned}
T2(J) &= T1*U(J), \\
T(0) &= EXP(T2(0)), \\
&\text{for } J = 1,2,\ldots,K, \\
T(J) &= \sum_{I=0}^{J-1} ((J-I)/J)*T(I)*T2(J-I) ,
\end{aligned}
$$

from (4.26).

The power function has been discussed in detail to point out, among other things, that functions which are easy to handle mathematically may require considerable care to program properly.

4. Projects for Chapter 4.

1°. The complete differential code list discussed in §1 can be formed by following each line of the function code list by the line or lines for the differential of the label of the line in the function code list, and then the resulting list can be packed, if necessary, to obtain a code list. For example, applied to (2.3), this gives the list

(4.79)

$$
\begin{aligned}
T1 &= X*Y \\
DT11 &= Y*DX \\
DT12 &= X*DY \\
DT1 &= DT11 + DT12 \\
T2 &= SIN(X) \\
DT21 &= COS(X) \\
DT2 &= DT21*DX \\
T3 &= T1 + T2 \\
DT3 &= DT1 + DT2 \\
T4 &= T3 + 4 \\
DT4 &= DT3 \\
T5 &= Y**2 \\
DT51 &= Y**1 \\
DT52 &= 2*DT51 \\
DT5 &= DT52*DY \\
T6 &= 3*T5 \\
DT6 &= 3*DT5
\end{aligned}
$$

$$T7 = T6 + 6$$
$$DT7 = DT6$$
$$F = T4*T7$$
$$DF1 = T4*DT7$$
$$DF2 = T7*DT4$$
$$DF = DF1 + DF2 \; ,$$

which can be packed to obtain a complete differential code list for F and DF. Note that (4.79) is obtained by <u>merging</u> the code list for F given in (2.3) with the list (3.40). Write a program to merge a code list with the list of its differentials and pack the result. Pack (4.79) and compare the results.

2°. Code the necessary subroutines based on Table III.1 and modify the coder of Project 3.4.1° (Project 1° of §4, Chapter 3) to produce subroutine call lists of the form (4.11) for function values and differentials.

3°. Code the subroutines from Tables IV.1 and IV.2, and modify the coder to produce subroutine call lists of the form (4.37).

4°. Code the exponentiation subroutines in §3, and compare the results obtained with those produced by the general formula (4.23) in the case one argument is a constant.

5°. A subroutine call list of the form (4.37) for the generation of Taylor coefficients can be executed in one of two ways: For a given K, one time through, or successively for K = 0, 1, 2, ..., until the desired number of Taylor coefficients are obtained. Code the latter method, and compare with the once-through execution of (4.37) to obtain, say, F(0), F(1), ..., F(10). The use of successive generation of Taylor coefficients as compared to a once-through generation may arise in problems in which the number of coefficients to obtain a given accuracy of approximation by the Taylor polynomial (4.5) is not known in advance, but must be computed from the result. This can occur, for example, in the numerical solution of differential equations by Taylor series methods [51], [53], [55], [80].

CHAPTER V

EXAMPLES OF SOFTWARE FOR AUTOMATIC DIFFERENTIATION AND
GENERATION OF TAYLOR COEFFICIENTS

In this chapter, a description will be given of computer programs for differentiation and generation of Taylor coefficients which have been written on the basis
of the methods described in the previous chapters. This software, developed over
a period of years at the Mathematics Research Center, University of Wisconsin-Madison,
is cited for illustrative purposes, and the description given here is based on personal knowledge and actual use of these programs. Thus, there is no claim of com-
pleteness made with regard to mention of software developed elsewhere for differentiation and series expansion; furthermore, it is not asserted that the programs described represent the state of the art. They do, however, demonstrate conclusively
that the ideas presented previously can be implemented as operational computer programs.

Historically, the development of the software described can be traced back to
the presentation in October, 1964, of the paper by R. E. Moore [51] to a conference
sponsored by the Mathematics Research Center. (The proceedings of the conference
were published as [66].) In his paper, Moore referred to a computer program [57]
developed under his direction for the solution of systems of ordinary differential
equations by Taylor series, in which the required Taylor coefficients were generated
automatically by use of formulas as given in Chapter 4 (see also [53], [55]). As a
result of hearing Moore's presentation, it occured to L. B. Rall that the function
code list used to obtain the sequence of calls to subroutines for the computation of
Taylor coefficients could also be used to produce code lists for the partial derivatives of the function, and that the resulting lists could be used for further differentiation as well as evaluation. The ability to obtain higher and mixed partial
derivatives in this way was then applied to the numerical solution of systems of
nonlinear equations [71], as will be considered in more detail later in connection
with the automation of Newton's method [25], [26], [39]. The task of programming
the technique of formula differentiation by the method of processing code lists as
described in Chapter 3 was given to Allen Reiter, who promptly produced a program
for the CDC 1604 [76]. R. E. Moore joined the Mathematics Research Center in Janu-
ary, 1965, and supervised the production by Reiter of programs for the generation of
Taylor coefficients [78] (see also [80]), and for interval arithmetic [77], [79].
The latter capability is needed in the automatic error estimation techniques developed by Moore for ordinary differential equations, and also for Rall's approach to

error estimation for approximate solutions of systems of nonlinear equations [25], [26], [39], [71], for which the additional coding was done by Julia H. Gray and D. Kuba. Reiter's original programs were extended and adapted to other machines, such as the CDC 3600 and the UNIVAC 1108/1110 by a number of workers, including Julia H. Gray, Dennis Kuba, and H. J. Wertz, whose contributions will be cited later.

Some of the earlier programs referred to above are essentially no longer operational, since they were written to a large extent in assembly language for machines which are now obsolete. The later versions, however, are mostly in FORTRAN, and can be used widely. This historical developement relates to the fact that machines were once much slower and had more limited storage than at present, and it was considered good programming practice to use the capabilities of the machine on hand to the maximum extent possible. Thus, since the CDC 1604 (and the CDC 3600) had 48-bit words, it was considered economical to pack as much information as possible into each word; for example, an entire line of a code list (label, entry, and perhaps a line number in addition to the label). Unfortunately, the coding required for packing and unpacking this information for processing is highly specific to the machine being used. The alternative of writing in a higher level language and consequently using a separate word for each item on a line in the code list (line label, label(s) of operand(s) and the operation) might have exhausted the available storage quickly on an older machine, and hence was rejected. Modern programming practice, on the other hand, is to make as much use of the capabilities of the language being used for programming as possible, with considerations of machine speed and storage being, to a certain extent, secondary. The more recent versions of the software described in this chapter reflect the latter point of view, and are written almost entirely in FORTRAN. Since the present discussion is more concerned with the basic principles of construction of software for differentiation and series expansion than with the use of a given program for these purposes, descriptions of some of the original programs are still useful. In particular, flow charts and other descriptive information may be used in programming in whatever language is desirable for the system or applications at hand, and this language, of course, may be something other than FORTRAN. The basic principles apply also if one does have to be concerned about the speed and/or the storage capacity of the hardware being used. For example, the introduction of a different generation of computers may affect programming practices; minicomputer and microcomputer programming has to take into account many of the limitations encountered with the very early electronic computers with respect to size and speed. The introduction of parallel processors, on the other hand, opens entirely new vistas.

The key item of software for differentiation and Taylor series expansion is the coder, that is, the program which translates formulas such as (2.2) into function code lists of the form (2.3), or subroutine call lists (4.11) or (4.37). This is true independently of the programming language and techniques used, since the code

or subroutine call list for the function is used in the construction of both derivatives and Taylor series of the function translated from its formula. It follows
that writing a good coder should have first priority in a modular plan for producing
software of the type discussed in this book. Once the language is chosen and the
necessary decisions are made about labeling and storing (that is, identifying) the
code lists produced as the result of formula translation, as well as about the internal structure of the lists (labels, defined operations), and the needed subroutine
library is developed, then the programming for differentiating or other processing
of the code lists obtained by translation or otherwise can proceed very naturally and
easily.

The importance of the coder goes beyond the particular application of differentiation. Most obviously, the function code list could be a step in the generation
of machine code for the evaluation of the function. On a slighly higher level, the
function code list might be used in the evaluation of the function in some type of
arithmetic not available as a standard machine feature (this may even include ordinary "real" arithmetic, that is, floating point arithmetic). In this connection,
one might think of complex, fractional, or, as will be cited extensively later, interval arithmetic. This is similar to what is done in ordinary compilers; variables
and constants can be declared to be of a certain "type", and a line in the function
code list would then generate a call to the appropriate subroutine to perform the
correct operation on the type of quantities encountered in the entry in that line,
and then the label of the line would be assigned the type corresponding to the operation indicated.

The fact that the function code list can be used for several purposes makes it
important to separate the process of formula translation into a function code list
from the other operations to be performed. For example, one may wish to evaluate a
function $f = f(x,y)$ in real arithmetic, and then later evaluate some derivative of f
such as $\partial^5 f/\partial x^3 \partial y^2$ in interval arithmetic. For these purposes, it would be handy to
have the function code list for f available; if this list had been simply translated
into a subroutine call list for the purpose of evaluation of the function, then it
would have to be reconstructed for the subsequent differentiation and evaluation in
interval arithmetic. The preparation of the coder as a separate module of the software also simplifies the analysis needed to understand the programming required in
applications, in contrast to attempting to perform other operations by subroutine
calls as the code list is generated.

The use of the function code list permits the generalization of the idea of
"type" in several ways not immediately related to ordinary arithmetic. As will be
discussed later in connection with the work of Kedem [37], [95], one may declare
variables U, V, to be of type "Taylor"; that is, U, V are actually to be vectors of
order K + 1, with coefficients $U(J)$, $V(J)$, $J = 0,1,...,K$ interpreted as their Taylor
coefficients in expansions in terms of some real variable ϕ. Then, a reference in

the function code list to

(5.1)   T = U/V

would result in a call to the subroutine for forming the Taylor coefficients T(J),
J = 0,1,...,K, as described in Chapter 4 (see (4.24)), and the label T would be assigned the type "Taylor" as a consequence of this operation. Of course, one has the
possibility of operations with various types, for example, an evaluation

(5.2)   V = U(0) + U(1)*T + U(2)*T**2 + ... + U(K)*T**K

of a Taylor polynomial at a real value T would yield a real result V. Before dis-
cussing this situation further, some basic programs implementing the ideas in the
two previous chapters will be described.

     1. CODEX and SUPER-CODEX. As mentioned above, the first program written at
the University of Wisconsin-Madison to carry out the method described in Chapter 3
was prepared by Allen Reiter [76] for the CDC 1604, and was adapted to the CDC 3600
by Julia H. Gray [30]. The program CODEX was specifically designed for use in the
Mathematics Research Center program NEWTON [25], [26], which was written by Gray under the direction of L. B. Rall, and will be described in a subsequent chapter.
(CODEX and NEWTON are also described in the book by Rall [71], which gives samples
of output obtained from these programs.) The description of CODEX given here is
taken essentially from the report by Reiter and Gray [30]. SUPER-CODEX, which differs from CODEX in a number of ways, will be described at the end of this section,
based on material prepared by H. J. Wertz [89], [90]. Versions of SUPER-CODEX were
written for the CDC 3600 and CDC 6600 by Wertz, using Reiter's original program CODEX
as a model. SUPER-CODEX was adapted to the UNIVAC 1100 series by Julia H. Gray,
specifically to the UNIVAC 1108, and it is operational also on the UNIVAC 1110 and
1160.

     The abstract of the report [30] describing CODEX states:

     "In designing general programs for the solution of systems of nonlinear equa-
tions, for numerical integration, and for many other mathematical procedures, one is
confronted with the need for a generalized differentiation routine. CODEX is a program for the CDC 3600 designed to meet this need. The program reads the functions
in from cards, translates them into code which is used in the differentiation and
evaluation of the functions. Likewise the code resulting from differentiation of a
function may be used in further differentiation and evaluation."

     The program CODEX implements the ideas put forth in Chapter 3. In the above
quotation, the word "code" refers to what is called a "code list" in this text. In
the report [30] describing CODEX, the operation of formation of the function code
list, known here as "coding", is called "compiling the expression", since formulas
are defined to be of the form

(5.3)   variable name = expression $,

where "expression" denotes a FORTRAN-like arithmetic expression consisting of numbers, variable names, operation names, and parentheses, arranged in a meaningful sequence in the FORTRAN sense. The special symbol $ is used simply to denote the end of the formula. For processing by CODEX, formula (2.2) would be written as

(5.4)   F = (X * Y + SINF (X) + 4) * (3 * Y ** 2 + 6) $

where F is the variable name, and (X * Y + SINF (X) + 4) * (3 * Y ** 2 + 6) is the expression. (Some specific technical details about the operation of CODEX will be suppressed in the following; the main purpose here is to explain the construction of its coder and differentiator, since these were used in more recent programs, and can serve as guides for software in other languages.)

Formulas of the type (5.4) are read by a format-free input package known as subroutine RDNUM(SYMBOL), based on the one due to A. Hassitt [34], which reads non-blank characters until a blank is found. (Hence, the blanks in (5.4) are significant for CODEX, but inconsequential for the following discussion.) As symbols, the program accepts the following:

(a) Variable names, consisting of a sequence of up to three non-blank characters (alphanumeric), the first of which is a letter;

(b) Operation names, consisting of a sequence of four non-blank alphanumeric characters, the first of which is a letter, or one of the following: +, -, *, /, **;

(c) Numbers in the form of signed or unsigned fixed-point integers, floating point numbers with decimal points or expressed as FORTRAN E-format numbers (all numbers are stored internally as single-precision floating point numbers by the program);

(d) The special symbols =, $, and left (and right) parentheses.

The subroutine library for the original version of CODEX consisted only of the functions sine, cosine, exponential, (natural) logarithm, and arctangent, for which the operation names are SINF, COSF, EXPF, LOGF, and ATAN, respectively. As may be seen from the flow-charts given below for the coder and differentiator, additional library functions can be added easily to a program of this type.

Once a function code list has been formed by the coder (see the flow-charts in Table V.1 below for a description of the process used by CODEX), then either the operation of _differentiation_ (formation of the derivative code list) with respect to any variable whose name appears in the function code list, or the operation of _evaluation_ can be applied, where the value defined by the code list is computed, and the result is assigned to the variable name associated with the code list, that is, the label of its last line. In CODEX, this is done by executing the sequence of instructions in the code list interpretively (see Tables V.3 and V.4 below). Once a code list has been formed, whether by the coder or the differentiator (flow-charts are given for this in Table V.2 below), then it can be differentiated further, evaluated, or printed out as a simple code list in the form shown in [71]. In order to simplify the discussion, these various functions of the program will be considered separately.

1.1. The coder. This portion of CODEX is particularly important for the reasons given at the beginning of this chapter, and because it also forms the basis of the programs SUPER-CODEX and TAYLOR, the latter to be described later in connection with the automatic generation of Taylor coefficients [27], [78]. In CODEX, the coder is referred to as subroutine READF, and is invoked by

(5.5)   CALL READF(NAMEF),

where NAMEF is the variable with value equal to the name of the formula to be read from cards. Thus, the function code list for (5.4) would have

(5.6)   NAMEF = F .

(There is also an initialization subroutine in CODEX, INIT(DUMMY), which must be executed before the formation of any function code lists; for the present discussion, this is an unnecessary technicality.)

The coder requires the assignment of various storage areas to code lists and other tables required in the translation of the formula and, of course, a table lookup subroutine. In CODEX, this is the subroutine SEARCH(NENTRIES,TABLE, ARGUMENT). Here, NENTRIES is the size of the table to be searched, TABLE is the name of the table (location of first element), and ARGUMENT is the object sought. The statement

(5.7)   N = SEARCH(NENTRIES,TABLE,ARGUMENT)

will return N = 0 if ARGUMENT is not in TABLE, otherwise, the value of N will give the position of ARGUMENT in TABLE. In CODEX, this subroutine is written in assembly language, and will not be discussed further here.

Some storage locations needed by the coder will now be defined in the way they are referred to in the following flow-charts. In CODEX, these locations are set aside as labeled COMMON blocks. As is customary, each set of storage locations is provided with a counter, which gives the number of items actually stored in the given tables at any time. The first set of storage assignments is

(5.8)   /FNAME/   K,LISTNAME(N),LISTAR(N),

where N is some suitably chosen unsigned integer giving the maximum size of the tables LISTNAME, which is the list of defined variable names (by formula or code list), and LISTSTAR is the starting addresses of their code lists (see below for further explanation). The number K is the list counter, which gives the number of items stored in LISTNAME (and consequently also in LISTSTAR). The code lists themselves are stored in

(5.9)   /COMP/   KC,LCOMOP(M),LCOMVAR(M),

where the names used in the flow-charts have been taken from [30]. The line in the code list being stored is assumed to be of the form

(5.10)   NDEST = LV1 (operation) LV2,

where, in the original version of CODEX, the operation was stored in the upper 24
bits of the word in LOCOMOP corresponding to the line (5.10), and the label of the
line NDEST (called the destination of the operation in [30]) is stored in the lower
24 bits. The label of the left variable LV1 is stored in the upper half of the corresponding word in LCOMVAR, which is set to blanks in the case of a unary operation,
and the label of the right variable LV2 is stored in the lower half of the same word.
On other systems, it may be more convenient to use four tables in place of (5.9),
that is, one for each element of the line (5.9) (label, operation, left variable,
right variable).

Another storage allocation is made to a table of constants,

(5.11)   /CONST/   KCON,CONTAB(L) ,

of suitable size. It has been found to be useful to use the first nine or so locations of this table for the positive integers 1,2,...,9, for example. Subsequent
locations in the table are then used for the storage of other constants in the order
in which they are encountered in the formulas being translated. In the code lists
produced by CODEX, constants are assigned names according to their position in the
table by a three-digit octal number followed by the letter C. If the convention for
storing small positive integers is followed as mentioned above, then 8 would be
called 010C in the code list produced by translation of a formula. (The form C010,
or, even better, C008 is preferable if the code list is to be processed later by a
FORTRAN-type compiler. The conventions about octal numbering and enumeration of
lines are given here merely to explain the programs CODEX and SUPER-CODEX and so output produced by these programs are understandable; the reader should regard them as
antiques, and not as patterns for modern software.)

A code list will usually consist of a number of lines preceding the line labeled
with the name of the formula being translated into the function code list corresponding to its expression. These are called intermediate lines (for example, the lines
labeled T1, T2, ..., T7 in the function code list (2.3) are intermediate lines), and
the values of their labels will occupy intermediate storage when the code list is
evaluated. The amount of this intermediate storage used at any time is entered into
the one-element common block

(5.12)   /KTST/   N .

This information is used to label lines in intermediate storage in the same way as
described above for constants, that is, a three-digit octal number followed by a letter, which will be T in this case. Thus, if 33 lines of intermediate code have been
formed in the process of differentiation and translation of one or more formulas,
then the next intermediate line will have the label 040T (again, T034 would be the
label consistent with present practice), and the entry in the common block KTST would
be increased to $34 = 40_8$. Another important storage allocation is

(5.13)   /LI/   LIBFUNCT(L) ,

which stores the list of operations recognized by the coder (that is, symbols for arithmetic operations and names of library functions).

In CODEX, the coder portion proper, or COMPILE, operates according to the flow-charts given in Table V.1 (Figures V.1a - V.1f) on the following pages. In order to make these charts more readable, the following definitions are given for the nomenclature used.

CONTAB - is the constants table.

LBANK - is the level of the current bank of operations. This level is moved up or down respectively by left and right parentheses.

LCOMOP and LCOMVAR - are the tables storing the code lists; the counter for these tables is KC.

LCOP - is the level of the operation being processed.

LISTOP - is the pushdown list of operations being processed.

LOPDE - is the last instruction in the LISTOP table (pushdown list).

LUB - indicates whether the current operation is binary or unary.

LVAT - is the pushdown list for the variables being processed.

LV1 - is the first (left) variable in the entry being created for the LCOMVAR list.

LV2 - is the second (right) variable in the entry being created for the LCOMVAR list.

NAMEF - is the name of the formula being processed.

NDEST - is the label of the line being created for the code list (in the terminology of [30], NDEST is the destination of the result of the operation currently being processed).

1.2. <u>The differentiator</u>. This portion of CODEX produces what have previously been called packed derivative code lists from function code lists prepared by use of the compilation subroutine (the coder), or by a previous differentiation. The differentiator is invoked by

(5.14)   CALL DIF(NAMEZ,NAMEF,NAMEX),

which differentiates the formula (code list) called NAMEF with respect to the variable called NAMEX; the resulting derivative code list is then given the name NAMEZ, which is assigned by the user. Thus, (5.6) followed by the formula (5.4) as data produces essentially the function code list (2.3); then,

(5.15)   CALL DIF(3HDXF,NAMEF,1HX)

would give the derivative code list for DXF = $\partial F/\partial X$ in a form similar to (3.41).

(In (5.15), the prefixes 3H and 1H are required by FORTRAN, since the subroutine DIF processes the <u>names</u> DXF, X rather than their values; a name with n characters would thus require the prefix nH. Also, in the code list resulting from (5.15), the intermediate variables designated by DXT2 and DXT4 in (3.41) will have octal designations such as 010T and 011T (that is, T8 and T9), since CODEX considers

TABLE V.1. FLOW-CHARTS FOR COMPILE

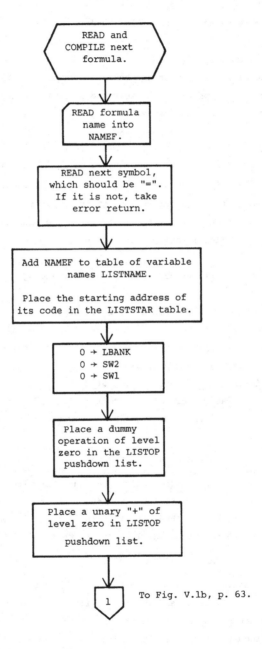

Figure V.la. Initialization for COMPILE.

TABLE V.1. (CONTINUED)

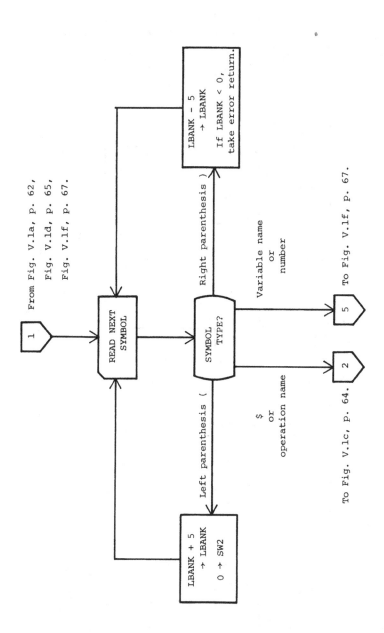

Figure V.1b. Read and Process Next Symbol.

TABLE V.1. (CONTINUED)

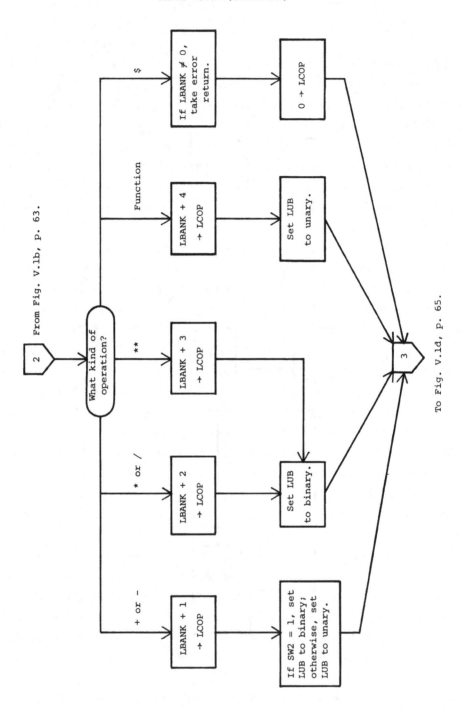

Figure V.lc. Operator Processing.

TABLE V.1.   (CONTINUED)

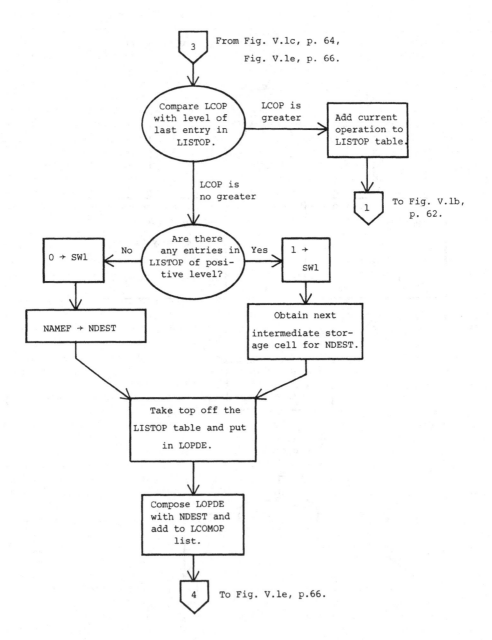

Figure V.1d.   Operator Processing (Continued).

TABLE V.1. (CONTINUED)

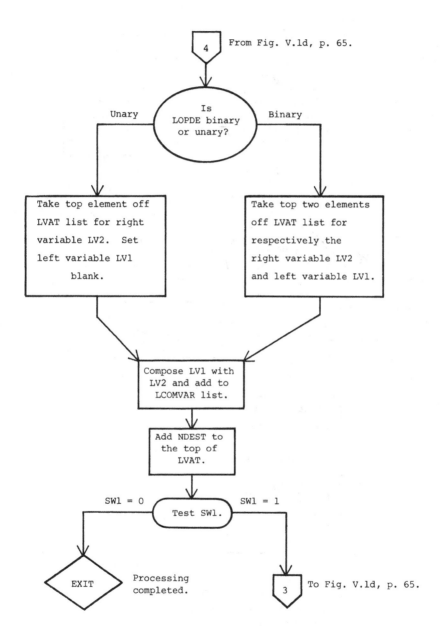

Figure V.1e. Operator Processing (Concluded).

TABLE V.1. (CONTINUED)

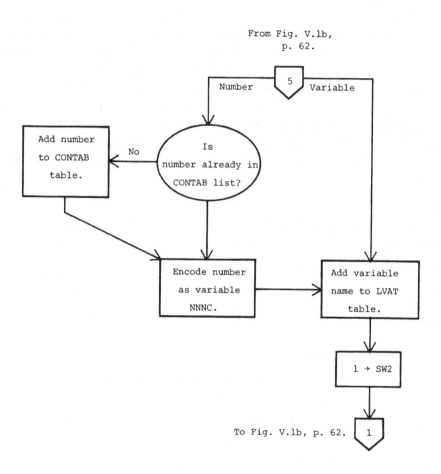

Figure V.1f. Variable Processing.

these simply as additional intermediate lines. These conventions are technicalities which are not essential to the understanding of the operation of the subroutine as shown in the flow-charts in Table V.2 below.)

The subroutine DIF requires the dependency tables LXY(I), LNZ(I), and the counter KXY, which are put into unlabeled (erasable) COMMON in CODEX. To explain the use of these tables, it is helpful to visualize the differentiation being performed as $Z = dY/dX$, where the variable (function) being differentiated has the name NY, it is differentiated with respect to the variable with name NX, and the result produced is named NZ. This process will result in entries in the table indicating the dependence of Y and Z on X. The dependency table is searched during the differentiation operation to determine if a derivative of a given label has been defined previously. If not, then the label is taken to represent a constant for the purpose of forming the derivative code list, and a partial derivative is obtained; otherwise, the result is the semi-total derivative. To illustrate this, consider the example of the function G defined by the formulas (3.6). If the subroutine call (omitting the Hollerith indicators nH)

(5.16)   CALL DIF(DXG,G,X)

is made in a CODEX program before U or V has been differentiated with respect to X, then the result will be the formal partial derivative

(5.17)   DXG = 2*X .

However, the sequence

(5.18)   CALL DIF(DXV,V,X)
                          CALL DIF(DXG,G,X)

will result in the semi-total derivative

(5.19) $DG/\partial X = \partial G/\partial X + (\partial G/\partial V)*(\partial V/\partial X) = 2*X + (EXP(V))*(U/Y)$,

since the information about the dependency of V on X is stored in the dependency table during the formation of the code list for $DXV = \partial V/\partial X$. Finally, the sequence

                          CALL DIF(DXU,U,X)
(5.20)   CALL DIF(DXV,V,X)
                          CALL DIF(DXG,G,X)

produces

(5.21) $DG/dx = \partial G/\partial X + (\partial G/\partial V)*(\partial V/\partial X) + (\partial G/\partial U)*(\partial U/\partial X)$.

Thus, CODEX is capable of producing partial, semi-total, or total derivatives of functions defined by an arbitrary sequence of formulas, as long as the user remembers to differentiate all relevant intermediate formulas before forming the final result desired. The price paid is a perhaps lengthy complete search of the dependency table LXY for the dependency of NY on NX when processing each line of the input code list. (The label NZ of dY/dX and its location in the code list are stored in LNZ.)

As in the case of COMPILE, various abbreviations are used in the flow-charts for DIF given in Table V.2 on the following pages. The relevant definitions are:

LDL – is the name of the derivative of the left variable of the line of the code list (simple formula) being differentiated.

LDR – is the name of the derivative of the right variable of the simple formula being differentiated.

LEDF – is the name of the derivative (being created) of the simple formula being differentiated.

LEF – is the name of the simple formula being differentiated.

LLF – is the name of the left variable in the simple formula being differentiated.

LLT – is the name of the left variable in the last entry into the code list tables currently being created.

LOP – is the operation of the simple formula being differentiated.

LRF – is the name of the right variable in the last entry into the code list tables currently being created.

ND – is the counter showing which simple formula (line in the code list being differentiated) is to be looked at next.

In the previous glossary for Table V.1, LLF and LRF were referred to as LV1 and LV2, respectively, and NDEST (the label of the line of the code list formed by the subroutine COMPILE) is LEF here; that is, the line being processed by DIF is assumed to be of the form

$$(5.22) \qquad LEF = LLF(LOP)LRF \ .$$

In the differentiation of (5.22), the subroutine DIF of CODEX forms one or more lines (simple formulas) with the structure

$$(5.23) \qquad LABEL = LLT(operation)LRT \ ,$$

where LABEL is either an intermediate line label nnnT (referred to as Ti, i a positive integer, in the flow-charts), or the final label LEDF of the derivative of the line being differentiated, which will be stored in the table LNZ giving names of derivatives corresponding to the labels of the corresponding line in the code list being differentiated.

The abbreviations NX, NY, NZ are also used in Table V.2 to denote the names in the subroutine call (5.14) in the standard form

$$(5.24) \qquad CALL \ DIF(NZ,NY,NX) \ ,$$

that is, the list known as NY is to be differentiated with respect to the name NX, and the resulting derivative code list is identified by the name NZ.

The subroutine DIF thus produces a form of what has been called a packed derivative code list in the previous chapters. An examination of Figure V.2j shows that rather than packing the list after it is produced, as suggested in Chapter 3, the

TABLE V.2. FLOW-CHARTS FOR DIF

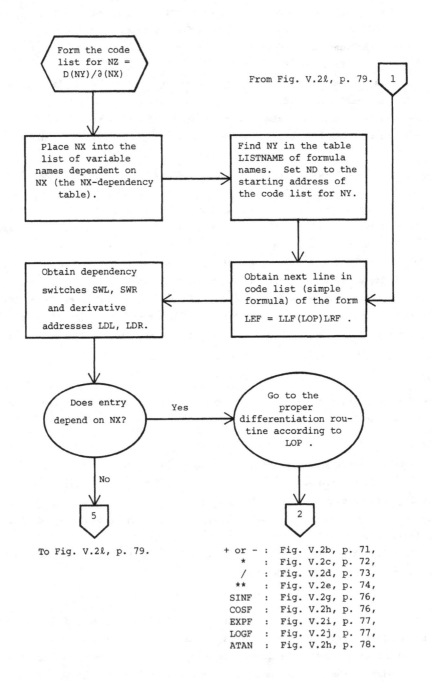

Form the code
list for NZ =
D(NY)/∂(NX)

From Fig. V.2ℓ, p. 79. 1

Place NX into the
list of variable
names dependent on
NX (the NX-dependency
table).

Find NY in the table
LISTNAME of formula
names. Set ND to the
starting address of
the code list for NY.

Obtain dependency
switches SWL, SWR
and derivative
addresses LDL, LDR.

Obtain next line in
code list (simple
formula) of the form
LEF = LLF(LOP)LRF .

Does entry
depend on NX?   Yes   Go to the
                                     proper
                                     differentiation rou-
                                     tine according to
                                     LOP .

No

5   2

To Fig. V.2ℓ, p. 79.   + or - : Fig. V.2b, p. 71,
                                   * : Fig. V.2c, p. 72,
                                   / : Fig. V.2d, p. 73,
                                  ** : Fig. V.2e, p. 74,
                                SINF : Fig. V.2g, p. 76,
                                COSF : Fig. V.2h, p. 76,
                                EXPF : Fig. V.2i, p. 77,
                                LOGF : Fig. V.2j, p. 77,
                                ATAN : Fig. V.2h, p. 78.

Figure V.2a. Initialization and Acquisition of a Simple Formula.

TABLE V.2. (CONTINUED)

Addition or Subtraction (+ or -) [2] From Fig. V.2a, p. 70.

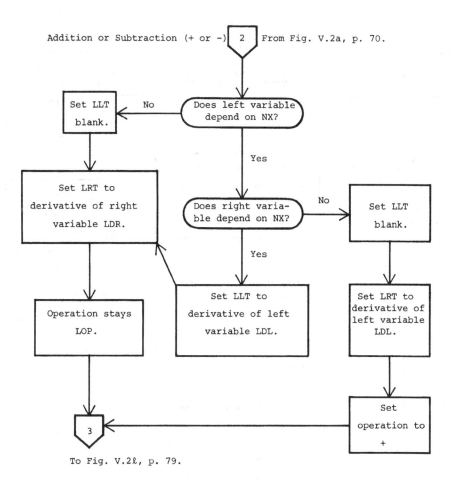

To Fig. V.2ℓ, p. 79.

Figure V.2b. Differentiation of Addition and Subtraction.

TABLE V.2. (CONTINUED)

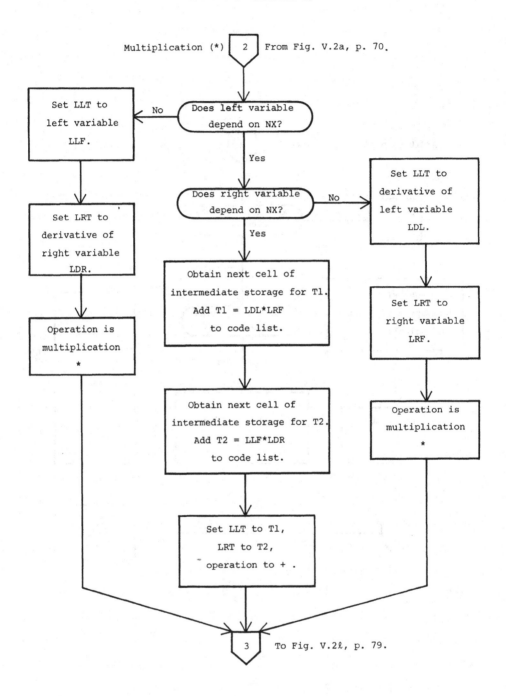

Figure V.2c. Differentiation of Multiplication.

73

TABLE V.2. (CONTINUED)

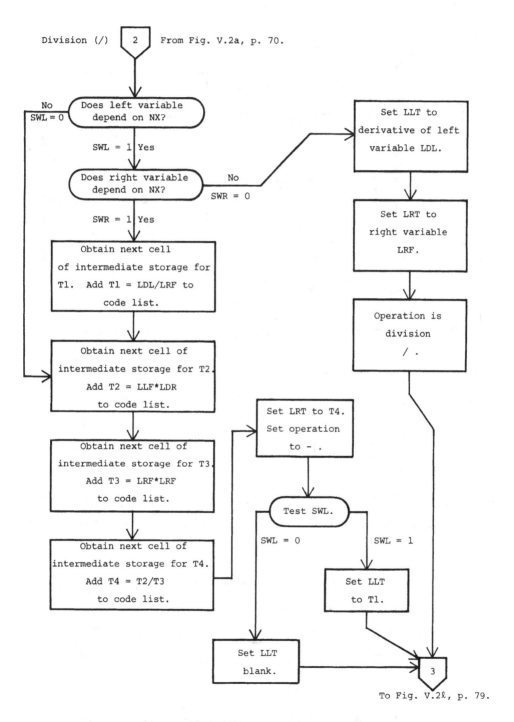

Figure V.2d. Differentiation of Division.

TABLE V.2.   (CONTINUED)

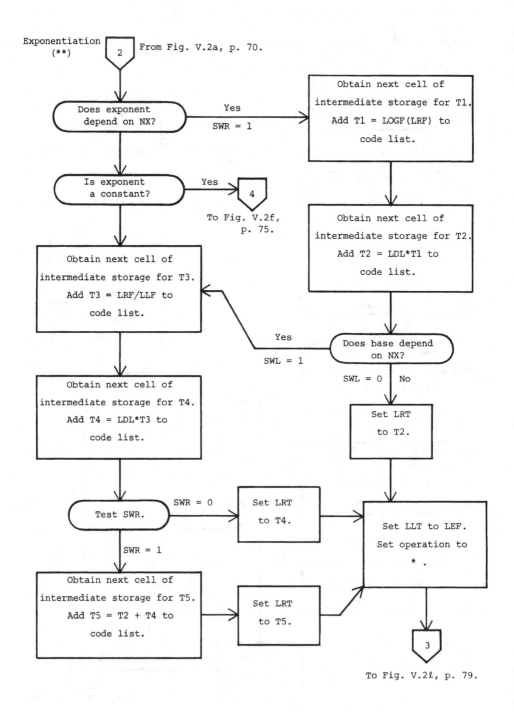

Figure V.2e. Differentiation of Exponentiation (Variable Exponent).

75

TABLE V.2. (CONTINUED)

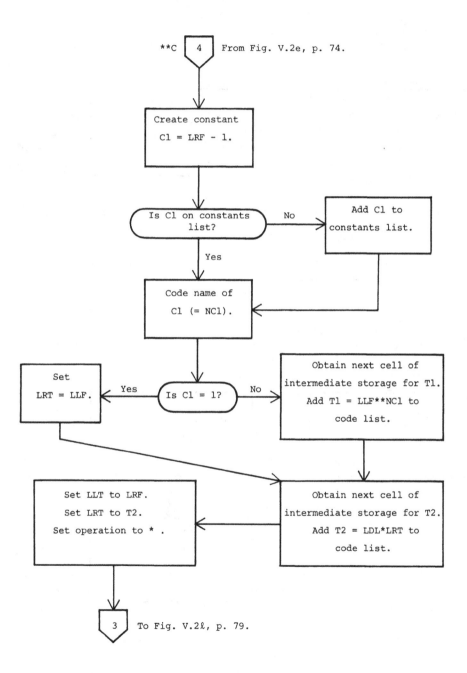

Figure V.2f. Differentiation of Exponentiation (Constant Exponent).

76

TABLE V.2. (CONTINUED)

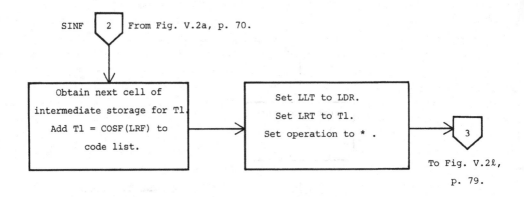

Figure V.2g. Differentiation of Sine.

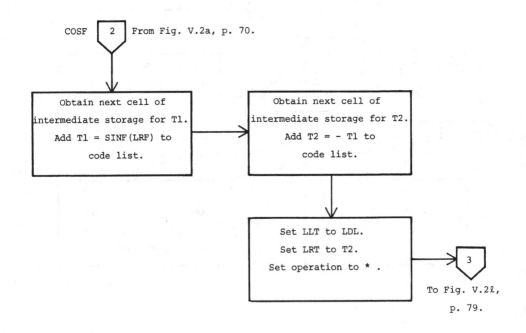

Figure V.2h. Differentiation of Cosine.

TABLE V.2.   (CONTINUED)

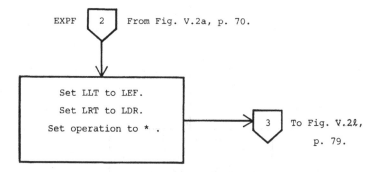

Figure V.2i.   Differentiation of the Exponential Function.

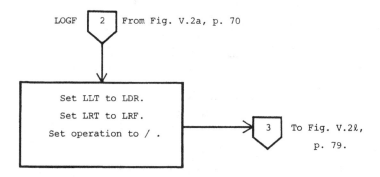

Figure V.2j.   Differentiation of the Logarithmic Function.

TABLE V.2.   (CONTINUED)

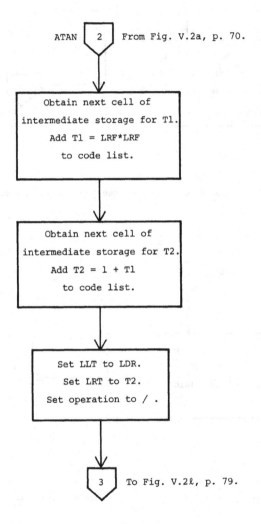

ATAN 2   From Fig. V.2a, p. 70.

Obtain next cell of
intermediate storage for T1.
Add T1 = LRF*LRF
to code list.

Obtain next cell of
intermediate storage for T2.
Add T2 = 1 + T1
to code list.

Set LLT to LDR.
Set LRT to T2.
Set operation to / .

3   To Fig. V.2ℓ, p. 79.

Figure V.2k. Differentiation of Arctangent.

TABLE V.2.   (CONTINUED)

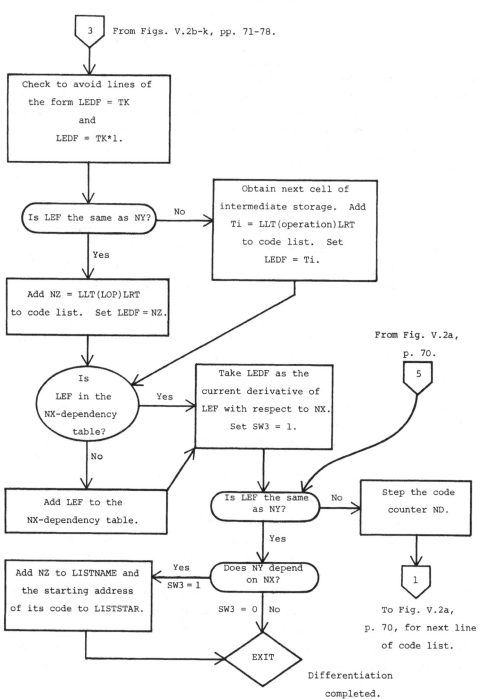

Figure V.2ℓ.   Formation of Packed Derivative Code List.

packing performed by DIF is done "on the fly"; that is, <u>trivial</u> code lines

(5.25) $$LEDF = TK \quad \text{or} \quad LEDF = TK*1$$

are simply not added to the code list being formed by this processor.

    1.3. <u>Other CODEX subroutines: ASSIGN, EVAL, and PRINT</u>. In order to do numerical computations with the code list representations of functions and their derivatives produced by COMPILE and DIF, an interpretive method is used. The interpretation of a function code list requires that all variables and parameters appearing in the original formula for the function from which the function or derivative code list was obtained be assigned specific numerical values. In CODEX, these values will be stored in the table (labeled COMMON block)

(5.26) $$\text{/VALUES/} \quad V(N) \ .$$

The counter for this table, K, is the same as for the /FNAME/ lists LISTNAME(N) and LISTSTAR(N). Thus, if F is the third name in LISTNAME of variables, then its value, when assigned or computed, will be stored as the third entry in the V list, which is called the <u>named variable</u> list. The values of intermediate storage variables are placed in a table called TSTORE in unlabeled (erasable) COMMON, and hence are written over the dependency tables LXY and LNZ.

    To give the numerical value VALUEX to the named variable NAMEX, that is, to store VALUEX in the table (5.26) in the position corresponding to NAMEX in the LISTNAME table, the following subroutine call is used:

(5.27) $$\text{CALL ASSIGN(NAMEX,VALUEX)} \ .$$

Normally, this will be used for independent variables, such as X, Y in (2.2), but it could also be used to assign a value to the dependent variable F arbitrarily, which could be used in the evaluation of a subsequent quantity dependent on F.

    In order to evaluate a function or derivative defined by a code list produced by COMPILE or DIF, an interpreter is invoked by the statement:

(5.28) $$VALUEF = EVAL(NAMEF) \ .$$

Here NAMEF occurs in the list of named variables LISTNAME. The value of VALUEF, obtained by interpretive execution of the code list for NAMEF, is stored in the table of values (5.26) in the position corresponding to NAMEF. All variable names appearing in the code list to be evaluated must have well-defined current values at the time EVAL is executed; that is, they must have previously appeared as parameters for either ASSIGN or EVAL. In addition, if a formula has been defined as a derivative of another formula, then the latter must be evaluated before the former, so that all intermediate variables in the code list for the first function will be given correct values. Thus, if A is the name of a formula (function code list), and B is produced by

(5.29) $$\text{CALL DIF(1HB,1HA,1HX)} \ ,$$

that is, $B = \partial A/\partial X$, then the sequence

(5.30)
$$VA = EVAL(1HA)$$
$$VB = EVAL(1HB)$$

must be used to obtain the current value of B, even if the value of A is not of interest, since the code list for B will be a derivative code list (see, for example, (3.41)), which will in general refer to named and intermediate variables in the function code list for A.

Flow charts for ASSIGN and EVAL are given in Figure V.3 and Figure V.4, respectively, on the following pages.

Another subroutine in CODEX is PRINT, which will output the code list designated by NAMEF,

(5.31)   CALL PRINT(NAMEF) .

The format of the printed output is a function code list, with lines of the form

(5.32)   vvvv = vvvv operation name vvvv,

for binary operations, or

(5.33)   vvvv = operation name (vvvv)

for unary operations. The symbols vvvv denote a named variable, an intermediate variable of the form nnnT, where nnn is an octal number, or a constant nnnC, with nnn again being the position of the constant in the CONTAB table in octal. Examples of the output of CODEX (which was written originally for the program NEWTON [25]) may be found in Chapter 4 of the book by Rall [71]. The operation name printed in (5.32) or (5.33) is, of course, one of the arithmetic operations or library functions permitted, respectively.

1.4. Features of SUPER-CODEX. The program SUPER-CODEX [89], [90] is operational on the CDC 6600 and the UNIVAC 1100 series (1108/1110/1160), and is based on CODEX. It provides a few additional features, among which are some additional library functions, additional instructions available to the user, and the option to obtain output either in the form of a code list or as a FORTRAN-like formula. This latter output format is produced by a compressor, which reads down the code list, inserts parentheses where necessary, and produces a linear formula as output. In the terminology of SUPER-CODEX, $LIST NAMEF will output the code list, while $DUMP NAMEF will produce the compressed formula for the function identified by NAMEF. This is illustrated by an example, namely, the function F defined by (2.2) and its derivative (3.41) with respect to X. The actual SUPER-CODEX processing gives:

(5.34) $COMPILE

          F = (X*Y + SIN(X) + 4) * (3*Y**2 + 6)

          $LIST F

     CODE LIST FOR F

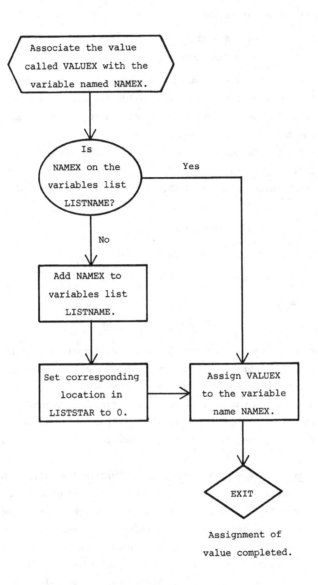

Figure V.3. Flow-chart for ASSIGN.

83

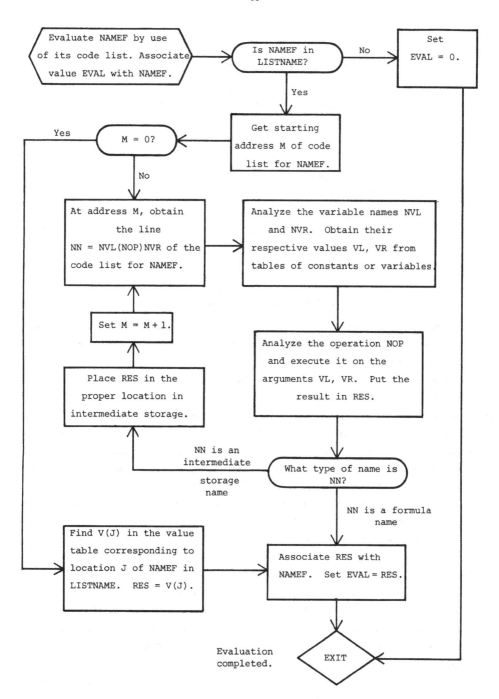

Figure V.4. Flow-chart for EVAL.

```
T00001 = X        *        ( Y       )
T00002 =          SIN      ( X       )
T00003 = T00001 +          ( T00002 )
T00004 = T00003 +          ( C00004 )
T00005 = Y        **       ( C00002 )
T00006 = C00003 *          ( T00005 )
T00007 = T00006 +          ( C00006 )
F      = T00004 *          ( T00007 )
```

$DIF F X DFDX

$LIST DFDX

CODE LIST FOR DFDX

```
T00010 =          COS      ( X       )
T00011 = Y        +        ( T00010 )
DFDX   = T00011 *          ( T00007 )
```

$DUMP DFDX DFDX

DFDX=(Y+COS(X))*(3.0*(Y**2)+6.0)

   This example was computed using SUPER-CODEX by Mr. Jørgen Wiene Owesen of the
Computer Science Department, University of Copenhagen. Note that the numeration is
in octal, not decimal.

   An examination of (5.34) reveals that the compressed formula for DFDX formed by
the SUPER-CODEX instruction $DUMP is produced by starting at the top of the code list
and working downward. (It is also possible to compress a code list into a formula
by starting at the bottom; this would be done by what could be called a walk-back,
or bottom-up compressor, in contrast to the top-down compressor used in SUPER-CODEX.)
The idea behind the introduction of the compressor was not to produce readable formulas for the user, but rather expressions which a FORTRAN compiler would use to produce more efficient code than obtainable from a code list. In computers of the type
for which SUPER-CODEX was written, the large number of references to storage required
by the straightforward compilation of a code list into machine code results in a less
efficient program than the compilation of a more complicated formula, such as (2.2),
in which the operations and values needed can be stored in high-speed registers.

   In fact, it was assumed at the time that SUPER-CODEX was written that the reason
for the lack of speed of CODEX was due to the interpretive execution of the code
lists produced by its coder and differentiator. However, a close analysis shows that
the most time-consuming segment of the differentiator in CODEX is the search of the
dependency table, a feature which is shared by SUPER-CODEX. Some alternative procedures, which would result in considerable gains in speed, will be presented as
projects at the end of the chapter.

   In addition to the library functions available in CODEX, SUPER-CODEX provides the
following functions to the user:

   ACOS - arccosine, ASIN - arcsine, ABS - absolute value, SQRT - square root, and the following special functions:

PLEGN(n,m,X) - the Legendre polynomial $P_{n,m}(X)$, where n is N + i or i, m is M
or j, M, N are integer variables, i, j are positive integer constants, and
X is a single variable;

BESSEL(n,X) - the Bessel function $J_n(X)$, where n is N + i or i, X is a single
variable;

ATAN2(x,Y) - is the two-variable arctangent used in conversion from rectangular
to polar coordinates, where x is a (FORTRAN) expression, and Y is a single
variable;

CDFN(x) - the standard normal cumulative distribution function with mean 0 and
variance 1.

The instructions available to the user of SUPER-CODEX will now be listed. Where
blanks are significant, they will be denoted by ^{b}; that is, X^bY means X Y. The
instructions $COMPILE, $DIF, and $LIST are similar to the corresponding ones in CODEX,
as can be seen from the example (5.34). SUPER-CODEX is written in such a way that
each instruction continues to process data following it until a new instruction is
encountered.

$COMPILE^{b}name = expression - forms code lists as in CODEX;

$DIF^{b}name$_1$^{b}name$_2$^{b}name$_3$ - differentiates the function called name$_1$ with respect
to the variable called name$_2$, and identifies the result as name$_3$, The
same conventions with regard to order in the formation of partial or semi-
total derivatives apply as in CODEX.

$DIFALDMP^{b}name$_1$^{b}name$_2^{b}var_1^{b}var_2$^{b}...var$_n^{b} - provides a way to form the Jacobian
matrix of the functions previously compiled or differentiated from name$_1$
to name$_2$, inclusive, in alphabetic order, with respect to the variables
var$_1$, var$_2$, ..., var$_n$ in turn.

To illustrate this last instruction, suppose that f_1, f_2, ..., f_m are functions
of the n variables x_1, x_2, ..., x_n. The <u>Jacobian matrix</u> of this system is the m×n
matrix J of partial derivatives

(5.35)
$$J = (\partial f_i / \partial x_j) = \begin{bmatrix} \dfrac{\partial f_1}{\partial x_1} & \dfrac{\partial f_2}{\partial x_2} & \cdots & \dfrac{\partial f_1}{\partial x_n} \\ \cdots & \cdots & \cdots & \cdots \\ \dfrac{\partial f_m}{\partial x_1} & \dfrac{\partial f_m}{\partial x_2} & \cdots & \dfrac{\partial f_m}{\partial x_n} \end{bmatrix}.$$

(If the Fréchet or Gâteaux derivative of the mapping from the n-dimensional vector
space R^n into the space R^m defined by the function $f = (f_1, f_2, ..., f_m)$ exists, that
is, for $f(x) = (f_1(x), f_2(x), ..., f_m(x))$, $x = (x_1, x_2, ..., x_n)$, then it is represented
by the matrix (5.35), which is a <u>linear</u> operator from R^n into R^m for fixed x [62],
[71]. The instruction $DIFALDMP constructs code for (5.35) by <u>columns</u>, with the

results stored in an array STORE(K); that is, $COMPILE Fl F2 ... FM followed by
$DIFALDMP Fl FM Xl X2 ... XN $ will result in STORE(1) = $\partial Fl/\partial Xl$, STORE(2) = $\partial F2/\partial Xl$,
... . Further SUPER-CODEX instructions are:

$DIFDMPNZ - performs the same operations as $DIFALDMP, except that derivatives
which are identically zero are not entered in the STORE(K) array.

$COPY - reads the following cards and writes them verbatum on the output tape
until a $ is found in column one, that is, until the next SUPER-CODEX in-
struction is encountered. This is useful in the construction of FORTRAN
programs; see the example at the end of this section.

$DUMP ^{b}name$_1$ ^{b}name$_2$ - writes all code generated for expressions named from name$_1$
to name$_2$ (in alphabetic order; see $DIFALDMP, for example, about the sig-
nificance of the delimiting names name$_1$ and name$_2$) on tape as FORTRAN expressions.

$LIST ^{b}name$_1$ ^{b}name$_2$ - outputs the code lists generated for all expression named
from name$_1$ to name$_2$ in alphabetic order.

$INIT - an initialization instruction, which resets all counters and lists, in-
cluding the dependency tables.

$C - indicates comments which will simply be copied onto the listing of the run-
stream, but which will not appear in the output of SUPER-CODEX.

$FOR - is changed into @FOR in the SUPER-CODEX output to invoke the FORTRAN
compiler to process output code for derivatives into machine code. See the
example at the end of this section.

$PUNCH - if the name TAPE11 (the output file for SUPER-CODEX) appears on the
card with the $PUNCH instruction, then the entire contents of TAPE11 will
be punched out after a $END card is encountered. If the remaining columns
of the $PUNCH card are blank instead, then a card will be punched for each
line of output produced on TAPE11 by the $DUMP, $DIFALDMP, and $DIFDMPNZ
instructions.

$END - terminates SUPER-CODEX operations.

The following example, prepared by Mr. Owesen, illustrates the production of
a FORTRAN subroutine using SUPER-CODEX instructions given above.

```
$COPY
$FOR   ,FIL.F
      REAL FUNCTION F(X)
$COMPILE
      F = SIN(X)*2 + X**2 + 1/X
$DUMP F F
   F = (SIN(X))*2.0+X**2+1.0/X

$COPY
      RETURN
      END
```

```
$FOR   ,FIL.DFDX
       REAL FUNCTION DFDX(X)
$DIF F X DFDX
$DUMP DFDX DFDX
   DFDX=(COS(X))*2.0+2.0*X-(1.0/(X*X))

$COPY
       RETURN
       END
$END
```

The output of the above SUPER-CODEX program consisted of the following:

```
 1.    @FOR   ,FIL.F
 2.           REAL FUNCTION F(X)
 3.           F=(SIN(X))*2.0+X**2+1.0/X
 4.           RETURN
 5.           END
 6.    @FOR   ,FIL.DFDX
 7.           REAL FUNCTION DFDX(X)
 8.           DFDX=((COS(X))*2.0+2.0*X-(1.0/(X*X)))
 9.           RETURN
10.           END
```

This can be compiled directly into FORTRAN subroutines for the function F and its derivative DFDX = dF/dX.

2. <u>TAYLOR and TAYLOR-GRADIENT</u>. These programs implement the method of generation of Taylor coefficients given in Chapter 4. The original program TAYLOR, written by Allen Reiter in 1965 [78], [80], employs the same coder as CODEX. After the set of functions being considered is read (since the main application of this program was to the numerical solution of systems of ordinary differential equations, the input section was written to handle a number of formulas), the function code lists obtained were used to generate <u>machine code</u> (for the CDC 1604) to execute the generation of Taylor coefficients line by line according to the formulas given in Chapter 4. When the values of the Taylor coefficients were to be computed, this machine code was executed to obtain the desired results. The program TAYLOR consists of the following main subroutines:

a) COMPILE reads in the whole group of statements describing the system of functions and translates the FORTRAN type formulas on the cards into function code lists. This subroutine is normally called only once during the course of a given program.

b) ZCODEREL (or ZCODEINT) is the differentiating routine. (In these names, REL refers to real (floating-point) arithmetic, and INT to interval arithmetic. At the same time he wrote CODEX and TAYLOR, Reiter wrote INTERVAL, an interval package to support computations in interval arithmetic needed in the solution of ordinary differential equations by Moore's methods [51], [53], [55]. INTERVAL is described in [77], [79].) This subroutine scans the code lists created by COMPILE, and generates actual CDC 1604 machine language code for obtaining the pth Taylor coefficient of each input function as a function of p and the first p Taylor coefficients of the independent input variables. This subroutine is also normally called only once during the course of a program.

c) DECODREL (or DECODEINT) is the subroutine compiled by ZCODEREL (or ZCODEINT). This subroutine is typically called many times; it is called once for each value of p, and may operate on many sets of values of the input variables and their Taylor coefficients.

In more detail, one starts the execution with a call of the subroutine COMPILE:

(5.36)   CALL COMPILE(N,K) ,

where N is the total number of equation names which appear on the left sides of the formulas to be read. The user must supply four storage arrays: The Y array with dimension Y(K+1,N) for the input variables and their Taylor coefficients, the Q array with dimension Q(K+1,30) for <u>shorthand</u> variables (such as U, V in (3.6)), the D array with dimension D(K+1,N) for the dependent variables, and the T array with dimension T(K+1,100) for intermediate variables. Of course, the literal constants 30, 100 appearing in the above dimension allocations could be increased, if necessary, providing that storage is available. The next step in the execution of TAYLOR is

(5.37)   CALL ZCODEREL(Y,D,Q,T,KSTORE)

in real arithmetic, with KSTORE = K + 1, or

(5.38)   CALL ZCODEINT(Y,D,Q,T,KSTORE)

in interval arithmetic, with KSTORE = 2*(K + 1). The actual program including evaluation of the Taylor coefficients would follow, with their values being obtained by

(5.39)   CALL DECODREL(J)   (or DECODINT(J)) ,

where J is the order of the Taylor coefficient, $0 \leq J \leq K$. The following must be observed:

i)   The values of Y(j+1,i) must be well-defined for $0 \leq j \leq J$ and for $1 \leq i \leq N$;

ii) DECODEREL/DECODINT must be called for all values of J in sequence, starting with J = 0.

The basic principles involved in writing the TAYLOR program are similar to those described for CODEX and the methods given in Chapter 4 concerning the implementation of the generation of Taylor coefficients of functions represented by code lists. The discussion of TAYLOR will consequently not be carried further, and attention will be turned to TAYLOR-GRADIENT, a program which is operational on the UNIVAC 1100 series in conjunction with the precompiler AUGMENT [12], [13], [14], [15], [16]. A detailed description of TAYLOR-GRADIENT may be found in the report [95] by Gershon Kedem, who wrote this sophisticated program at the Mathematics Research Center, with the documentation appearing in 1976. The basic technique implemented for differentiation in this program is the replacement of the function code list by a sequence of calls to subroutines for differentiation or the calculation of Taylor coefficients as described in Chaper 4 (see (4.11) and (4.37)). Some features of TAYLOR-GRADIENT which are worthy of note are that, by the use of the AUGMENT precompiler, the user may declare variables (names) to be of TYPE TAYLOR or TYPE GRADIENT. If F is declared to be of TYPE TAYLOR, then it will be treated as the vector of order N + 1 of its Taylor coefficients of order up to and including N, for some prescribed N, that is,

$$(5.40) \qquad F = (F(0), F(1), \ldots, F(N-1), F(N))$$

as in Chapter 4. On the other hand, if G is a function of N variables X1, X2, ..., XN, and the partial derivatives $\partial G/\partial Xi$ are denoted by DGXi(X) at the point X = (X1, X2,...,XN), then if G is declared to be of TYPE GRADIENT, it will be treated as the vector of order N + 1

$$(5.41) \qquad G = (G(X), DGX1(X), \ldots, DGXN(X))$$

consisting of the value of the function G at X and the values of its N partial derivatives with respect to X1, X2, ..., XN at X.

The input to the TAYLOR-GRADIENT program consists of an entire FORTRAN program, rather than just a list of formulas. As the variables of TYPE GRADIENT or TYPE TAYLOR are encountered in the program, code lists for the functions involved are analyzed to produce calls to the required subroutines for derivatives or Taylor coefficients.

Because of the identity of first derivatives and first Taylor coefficients, the subroutines used to generate Taylor vectors (5.40) can also be used to obtain gradient vectors (more properly, <u>augmented</u> gradient vectors including the function value) of the form (5.41). Note that if one takes X1, ..., XN to be Taylor vectors, with X1 = (X1, 1, 0,..., 0) and Xi = (Xi, 0, 0,..., 0) for i = 2,3,...,N, then the Taylor coefficient

$$(5.42) \qquad G(1) = DGX1(X) ,$$

the second component of the augmented gradient vector (5.41), and, of course, one has G(0) = G(X). Now, letting X1 = (X1, 0, 0, ..., 0), X2 = (X2, 1, 0, ..., 0), and Xi = (Xi, 0, 0, ..., 0), i = 3,4,...,N, one gets

$$(5.43) \qquad G(1) = DXG2(X) ,$$

and so on. Thus the routine for Taylor coefficients may be used for the calculation of <u>first</u> partial derivatives.

This method for obtaining gradient vectors, while convenient, is somewhat wasteful in that the value of G(X) = G(0) is calculated N times, once for each partial derivative. As mentioned previously, the method of Taylor coefficients is also not suitable for the computation of higher partial derivatives by declaring the obtained components DXG1, DXG2, ..., DXGN of (5.41) also to be of TYPE GRADIENT. Since the use of TAYLOR-GRADIENT involves AUGMENT, lack of space prevents a detailed description here. For further details, including a microfiche listing of TAYLOR-GRADIENT, the report [95] should be consulted.

3. Projects for Chapter 5.

1°. Speed up the portion of the differentiator in CODEX which requires search of the dependency table. One method to investigate is to start the search with the last label entered, since entries are often labels from the previous line or nearby lines of the code list. Another method is to use a random search or storage of labels ("hashing"). A third approach could be based on a Boolean vector for each label, in which the ith entry is T if the label depends on the ith variable, and F otherwise. If this vector is denoted by B(label), and

(5.44)   label = left(op)right ,

then B(label) = B(left) + B(right), using Boolean addition.

Compare one or more of these methods with the CODEX differentiator in terms of speed, after coding both in your favorite language.

2°. The logic of the program TAYLOR, in which the subroutine call list is executed repeatedly in order to generate successive Taylor coefficients, is based on application to the initial-value problem

(5.45) $y' = f(x,y), \quad y(x_0) = y_0.$

For $X(0) = x_0$, $Y(0) = y_0$, formula (5.45) gives the Taylor coefficient

(5.46) $Y(1) = f(X(0),Y(0));$

this value and $X(1) = 1$ can be used to obtain $Y(2)$ by differentiation of f, and so on. The expansion of a given function $f(x)$, however, can be done either as above, or a fixed number of Taylor coefficients K can be obtained by once-through evaluation, using the appropriate subroutines, since $X(0) = x_0$, $X(1) = 1$, $X(J) = 0$ for $J = 2, 3, \ldots$ are known. Write a program of this type and compare with respect to speed with a program of TAYLOR type which uses successive generation of coefficients.

Note: In case it is not known in advance how many Taylor coefficients will be needed, a successive generator may be preferable to a once-through program.

Code both types in your favorite language.

CHAPTER VI

AUTOMATIC COMPUTATION OF GRADIENTS, JACOBIANS,

HESSIANS, AND APPLICATIONS TO OPTIMIZATION

The central ideas in this chapter refer back to the method presented in Chapter 3 for differentiation by processing code lists for functions and previously obtained derivatives. There are two aspects of the application of automatic differentiation to be considered: The actual performance of calculations requiring the evaluation of derivatives, such as in optimization and the solution of nonlinear systems of equations, and secondly, the application of mathematical theorems which require information obtainable from derivatives, such as Lipschitz constants, in the verification of their hypotheses. Emphasis will be on the first aspect in this chapter, particularly in connection with optimization, with the solution of equations and the second topic being deferred to subsequent chapters. Here, a description of the code lists needed will be given, and some indication of their usefulness in smooth unconstrained and constrained optimization problems.

1. Gradient vectors and code lists. As in Chapter 3, functions f of n real variables,

$$(6.1) \qquad f(x) = f(x_1, x_2, \ldots, x_n)$$

will be considered, where it is convenient to regard f also as a function of the single vector variable

$$(6.2) \qquad x = (x_1, x_2, \ldots, x_n)^T,$$

and f is taken to be defined at least on some domain $D \subset R^n$, where R^n is the n-dimensional real vector space. Since f is assumed to be real-valued, this information is abbreviated as $f: D \subset R^n \to R$ [62]. The (Fréchet or Gâteaux) derivative of f at x, if it exists [62], [71], is the row vector (linear functional on R^n),

$$(6.3) \qquad f'(x) = (\frac{\partial f(x)}{\partial x_1} \quad \frac{\partial f(x)}{\partial x_2} \quad \ldots \quad \frac{\partial f(x)}{\partial x_n})$$

that is, $f'(x): R^n \to R$ is a linear operator (functional) mapping R^n into R at each x for which it exists. The gradient of f at x is the transpose of $f'(x)$, and will be denoted by

$$(6.4) \qquad \nabla f(x) = f'(x)^T ,$$

and is a vector belonging to R^n considered as a space of column vectors; hence,

$\nabla f(x)$ is often called the <u>gradient vector</u> of f at x. Since the components of (6.3) and (6.4) are exactly the same, their evaluation (or construction of code lists or subroutines for their evaluation) will be referred to as a <u>gradient computation</u>, with the result taken to be the gradient (6.4) or the derivative (6.3) as desired for interpretation or subsequent computation. (In finite dimensional vector spaces, as well as in Hilbert spaces in general, vectors belonging to the space and linear functionals on the space may be identified. From a computational standpoint, however, row and column vectors may be stored (as well as used) in distinct fashions, and thus an identification might not be appropriate.)

It follows that a gradient computation may be performed by the method of code lists by first forming the function code list for F = f, using the coder, which can then be processed by the differentiator to obtain the derivative code lists for

(6.5) $\qquad\qquad DFDXi = \partial f/\partial x_i, \quad i = 1,2,\ldots,n.$

The sequence of n code lists:

(6.6)
$$\begin{array}{l} \text{Derivative code list for DFDX1} \\ \text{Derivative code list for DFDX2} \\ \cdots \quad \cdots \quad \cdots \quad \cdots \quad \cdots \quad \cdots \quad \cdots \\ \text{Derivative code list for DFDXn} \end{array}$$

will be called a <u>gradient code list</u> for F. Evaluation of the lists (6.6), assuming that F has been evaluated, will produce the components of the vectors (6.3) and (6.4) at the given point $X = (X_1, X_2, \ldots, X_n)^T$. Thus, the function code list for F and the gradient code list (6.6) may be used to evaluate a variable of TYPE GRADIENT as defined in the previous chapter (see (5.41)).

The code list (6.6) may be preceded by the code list for F and the resulting list packed to obtain a <u>complete</u> gradient code list, which would evaluate the components (6.5) of the vectors (6.3) and (6.4), but not the function F. However, in most applications, the value of the function F as well as its derivative or gradient is desired, so there is essentially no wasted effort in adjoining the code list for F to the list (6.6) for computation. The result will be called the <u>augmented</u> gradient code list for F.

The method of generation of Taylor coefficients can also be used to calculate derivatives or gradient vectors of a function F. Take F = (F(0),F(1)) to be of TYPE TAYLOR, where F(0) is the value of F at some point X, and F(1) = dF(X)/d$\emptyset$ is its first Taylor coefficient in its expansion in terms of the variable $\emptyset$. The components $X_1, X_2, \ldots, X_n$ of the vector X can also be taken to be of TYPE TAYLOR, that is, $X_i = (X_i(0), X_i(1))$, also in terms of expansion in the (conveniently fictitious) variable $\emptyset$. Taking

(6.7) $\qquad\qquad X_i(1) = \delta_{ij} = \begin{cases} 0 \text{ if } i \neq j, \\ 1 \text{ if } i = j, \end{cases}$

(δ_{ij} is called the <u>Kronecker delta</u>), one obtains

(6.8) $F(0) = F(X)$, $F(1) = DFDXi$

for $i = 1,2,\ldots,n$, and thus the result is a gradient computation. There is a certain
amount of waste connected with this method, however, since $F(X) = F(0)$ is computed
n times, which is n - 1 times more than necessary, and so the computation based on
the use of the code list for F, evaluated once, and the gradient code list (6.6) appears to be more efficient.

An example of an augmented gradient code list can be obtained from the example
(2.2). The augmented gradient code list for F is easily seen to be

$$
\begin{aligned}
T1 &= X*Y \\
T2 &= SIN(X) \\
T3 &= T1 + T2 \\
T4 &= T3 + 4 \\
T5 &= Y**2 \\
T6 &= 3*T5 \\
T7 &= T6 + 6 \\
F &= T4*T7 \\
T8 &= COS(X) \\
T9 &= Y + T8 \\
DFDX &= T9*T7 \\
T10 &= 2*Y \\
T11 &= 3*T5 \\
T12 &= T4*T11 \\
T13 &= X*T7 \\
DFDY &= T12 + T13 \ .
\end{aligned}
$$

(6.9)

In actual practice, coding of functions is often simplified by the use of what
Reiter [80] calls <u>shorthand variables</u> for expressions which occur several times in
the formula of interest, or are quantities which may be useful elsewhere in the program by entering into the expressions for several functions, or for other reasons.
In Chapter 3, (3.6) illustrates the introduction of the shorthand variables U, V in
coding the formula (3.5) for the function G. Shorthand variables used in coding the
formula for f do not enter explicitly into the derivative and gradient vectors (6.3)
and (6.4), respectively; hence, the partial derivatives appearing there are actually
semi-total derivatives. Suppose that the coding of F involves the shorthand variables V1, V2, ..., Vm, where V1 depends (for the purpose of the gradient computation)
on X, V2 may depend on X and V1, and, in general, Vk will depend on X, V1, V2, ...,
and V(k-1). Thus, in brief, the function code list for F would consist of the sequence of code lists for the shorthand variables, followed by the code list for the
expression of F in terms of X, V1, V2, ..., Vm, that is, the list,

$$\text{(6.10)} \qquad \begin{array}{l} \text{Code list for V1} \\ \text{Code list for V2} \\ \text{... ... ...} \\ \text{Code list for Vm} \\ \text{Code list for F .} \end{array}$$

Because of the order of the dependencies among the shorthand variables, the augmented gradient code list for F can be obtained by following (6.10) with

$$\text{(6.11)} \qquad \begin{array}{l} \text{Gradient code list for V1} \\ \text{Gradient code list for V2} \\ \text{... ... ...} \\ \text{Gradient code list for Vm} \\ \text{Gradient code list for F .} \end{array}$$

An alternative organization of the computation is

$$\text{(6.12)} \qquad \begin{array}{l} \text{Code list for V1} \\ \text{Gradient code list for V1} \\ \text{Code list for V2} \\ \text{Gradient code list for V2} \\ \text{... ... ...} \\ \text{Code list for Vm} \\ \text{Gradient code list for Vm} \\ \text{Code list for F} \\ \text{Gradient code list for F .} \end{array}$$

In (6.11) and (6.12), it is to be understood that each gradient code list is formed with respect to $X1, X2, ..., Xn$ as the variables of differentiation.

Although the combination (6.10)-(6.11) or the list (6.12) may appear to be complicated in comparison to (6.6), the results obtained by using shorthand variables may actually be much simpler than without, especially if one or more expressions appear a number of times in the formula for F expressed as a function of $X1, X2, ..., Xn$ only. In general, the judicious use of shorthand variables is good programming practice, and leads to overall efficiency in evaluation and differentiation of complicated expressions for functions. The introduction (and suppression) of shorthand variables will also be discussed below from a slightly different standpoint in connection with Jacobian matrices. Before that, a brief discussion of the application of gradient computations to optimization problems will be given.

2. <u>Gradients and optimization problems</u>. One of the most important problems in applied mathematics, particularly in areas having economic implications, is to find the maximum or minimum values (called <u>optimal</u> values) of a given functional $f: D \subset R^n \to R$. The definition of the domain D of f may involve conditions called <u>constraints</u> which must be satisfied. Such conditions are of fundamental importance in the field known as <u>mathematical programming</u>. For the present, we shall take $D = R^n$, in which

case the problem under consideration is known as an <u>unconstrained</u> optimization problem, and deal with constraints later. If f is differentiable, then it is well-known
that the (local) maxima and minima of f occur at what are called the <u>critical</u> <u>points</u>
of f, which are the points x° at which the gradient of f vanishes, that is,

(6.13) $$\nabla f(x°) = 0 ,$$

where 0 in (6.13) denotes the zero vector (or <u>origin</u>) of R^n, $0 = (0,0,\ldots,0)$. Critical points of f at which f does not assume a local maximum or minimum value are
called <u>stationary</u> <u>points</u> of f; these may also be of interest in certain applications.

Equation (6.13) expresses a condition on each of the n components of the gradient vector $\nabla f(x°) = (f_1, f_2, \ldots, f_n)$, where

(6.14) $$f_i(x_1, x_2, \ldots, x_n) = \frac{\partial f(x)}{\partial x_i} ,$$

$i = 1, 2, \ldots, n$. Finding critical points of the functional f is thus equivalent to
solving the (generally nonlinear) system of n equations in n unknowns,

(6.15)
$$f_1(x_1, x_2, \ldots, x_n) = 0,$$
$$f_2(x_1, x_2, \ldots, x_n) = 0,$$
$$\cdots \quad \cdots \quad \cdots \quad \cdots \quad \cdots$$
$$f_n(x_1, x_2, \ldots, x_n) = 0,$$

for $x = x° = (x_1°, x_2°, \ldots, x_n°)$. Solution of the system (6.15) is one of the fundamental
techniques in the solution of unconstrained optimization problems, and also may be
involved in certain methods for constrained problems.

By the use of software for automatic differentiation, the subroutines required
for the evaluation of the functions $f_i(x_1, x_2, \ldots, x_n)$, $i = 1, 2, \ldots, n$, appearing in
(6.15) may be obtained simply by constructing the gradient code list for f from its
defining formula (or formulas). In optimization problems, the value of f is of basic
importance, and is not redundant information produced as a by-product to the calculation of the gradient vector. As a matter of fact, sometimes only the <u>optimal</u> <u>value</u>
of f,

(6.16) $$f° = f(x_1°, x_2°, \ldots, x_n°)$$

is all that is sought, with the coordinates of the <u>optimal</u> <u>point</u> $x° = (x_1°, x_2°, \ldots, x_n°)$
being of no interest in themselves. In most applications, however, the values of f°
and x° are both significant.

Automatic differentiation can also be used in a type of <u>constrained</u> optimization problem, namely,

$$\text{(a) optimize } f = f(x_1, x_2, \ldots, x_n),$$

(6.17)   (b) subject to:

$$g_1(x_1, x_2, \ldots, x_n) = 0,$$

$$g_2(x_1, x_2, \ldots, x_n) = 0,$$

$$\ldots \quad \ldots \quad \ldots \quad \ldots \quad \ldots$$

$$g_m(x_1, x_2, \ldots, x_m) = 0.$$

If the <u>constraint functions</u> g_1, g_2, $\ldots$, g_m are differentiable, then a system of equations for critical points of this problem (which include all the optimal points) may be obtained by the method of <u>Lagrange multipliers</u> [83]. These are m new variables introduced by the user, which, in addition to x_1^o, x_2^o, $\ldots$, x_n^o, satisfy the system of n + m equations consisting of equations (6.17)(b) above, and the n equations

(6.18) $$f_i(x_1, x_2, \ldots, x_n) + \sum_{j=1}^{m} \lambda_j \cdot g_{j,i}(x_1, x_2, \ldots, x_n) = 0,$$

$i = 1, 2, \ldots, n$, where the Lagrange multipliers are denoted by λ_1, λ_2, $\ldots$, λ_m in (6.18). The optimal values λ_1^o, λ_2^o, $\ldots$, λ_m^o of the multipliers may be of interest in addition to the coordinates of the optimal point x^o and, of course, the optimal value f^o of f [83]. In (6.18), derivatives of g_1, g_2, $\ldots$, g_m are denoted by use of the convention

(6.19) $$g_{j,i}(x_1, x_2, \ldots, x_n) = \frac{\partial g_j(x)}{\partial x_i},$$

by analogy with (6.14). Thus, all the subroutines necessary for computation of the functions entering into the system of equations (6.17)(b) and (6.18) are obtainable from the gradient code lists for f and g_1, g_2, $\ldots$, g_m.

Software to form the system of equations (6.15) for unconstrained optimization automatically, or the system (6.17)(b) and (6.18) for constrained optimization in the case of smooth constraints would be highly desirable, as a number of problems in mechanics and other applications fall into this category. No programs of this type are described in the literature cited in the survey [74]; however, optimization software of this type is available commercially from PROSE, Inc. [84], according to information received since [74] was published.

The constraints (6.17)(b) are sometimes expressed in the form

(6.20) $$g_j(x_1, x_2, \ldots, x_m) = c_j, \quad j = 1, 2, \ldots, m,$$

which does not alter the equations (6.18), since the derivatives (6.19) of (6.20) have the same values as for (6.17)(b). The formulation (6.20), however, gives rise to the interpretation of the Lagrange multipliers λ_j as rates of change of the optimal value f^o of f with respect to the constants c_j in the constraints (6.20), that is,

(6.21) $$\lambda_j^o = \frac{\partial f^o}{\partial c_j}, \quad j = 1, 2, \ldots, m,$$

at the optimal point x = x° [83]. These values of the Lagrange multipliers may thus
be significant in certain applications in mathematical economics and other areas.
In other problems, however, the Lagrange multipliers are regarded simply as auxiliary
variables, introduced to aid in the solution of the constrained optimization problem
(6.17).

Once the system of equations for unconstrained or constrained optimization are
formed, automatically or otherwise, they must be solved to obtain the desired information about the problem. There are numerous methods for solving finite systems of
nonlinear equations, many of which can be found in standard books on the subject,
such as [62], [63], [65], [71]. Thus, the subroutines described above which form
the functions on the left sides of the equations of the system will provide input to
the program used to solve the system. One method for the solution of systems of nonlinear equations is Newton's method (sometimes also called the Newton-Raphson or the
Newton-Kantorovich method), which will be described in more detail in §3 of this chapter, also makes use of the derivatives of the functions forming the left sides of the
system of equations. If these functions have been prepared in the form of code lists
by a coder or differentiator, then the required derivatives can also be obtained automatically.

It should be noted that a common alternative to the use of derivatives in optimization and other problems is to approximate derivatives by finite differences,
for example,

$$(6.22) \qquad \frac{\partial f}{\partial x_i} \approx \frac{1}{h}[f(x_1,x_2,\ldots,x_i+h,\ldots,x_n) - f(x_1,x_2,\ldots,x_i,\ldots,x_n)],$$

i = 1,2,...,n, for some h > 0. Supposing that $f(x) = f(x_1,x_2,\ldots,x_n)$ has been evaluated, each of the approximations (6.22) requires one additional evaluation of the
function f. In the absence of software for differentiation of f, this use of finite
difference approximations is attractive, since coding the needed derivatives by
hand is a dull, routine task, and extremely susceptible to the introduction of error,
even for moderate values of n. Thus, methods based on the use of the approximations
(6.22) are widely used, and are effective in practice [21]. However, some of the
claims that have been made for finite difference methods as opposed to methods which
use derivatives may be misleading. First of all, if the gradient code list for f
has been compiled into efficient machine code, there should be little difference in
overall execution time between the computation of the approximate values (6.22) and
the derivatives $\partial f(x)/\partial x_i$ obtained by automatic differentiation of the formula for
f. (In many cases, of course, the functional f is defined by a subroutine instead
of a single formula, or set of formulas if shorthand variables are used. In this
situation, the TAYLOR-GRADIENT program is applicable [37], [95].)

Secondly, as is well-known to users of finite difference approximations to derivatives [21], there is a loss of significant digits in the finite-difference approximation due to cancellation in (6.22), the error being magnified absolutely by

multiplication by the large number 1/h. Thus, one has a choice between an inaccurate approximation to the derivative resulting from taking h too large, or a complete loss of significant digits in the difference $f(x_1,\ldots,x_i+h,\ldots,x_n) - f(x_1,\ldots,x_i,\ldots,x_n)$, followed by multiplication by the large number 1/h, if h is taken too small. The standard remedy is to take $h = \sqrt{\eta}$, where η is the absolute precision with which $f(x)$ can be calculated [21]. Thus, a somewhat inaccurate value of the derivative is the price paid to keep the computation meaningful. However, this type of inaccuracy is of no consequence in a number of applications. In connection with the use of interval arithmetic, on the other hand, the use of derivatives is usually preferable to finite differences, since, for example,

$$(6.23) \qquad\qquad [0,1] - [0,1] = [-1,1],$$

so that cancellation of significant digits is replaced by widening of intervals, and is aggravated still more by multiplying by the large number 1/h.

Another advantage of the use of derivatives in certain circumstances is that it is possible to apply relevant theorems directly concerning convergence, truncation error, or other properties of the results being computed without having to take into account the additional error of approximation of derivatives by finite differences. Examples of the use of derivatives in this fashion will be given in later chapters in connection with the convergence of Newton's method and the error of numerical integration.

The main point is, given the availability of software for automatic differentiation, one should not make a Shibboleth of using or avoiding derivatives in all circumstances. The choice between derivatives and finite difference approximations should be made on whatever analysis (and practical experience) is available concerning the particular type of problem actually being solved computationally. The person who would avoid the use of derivatives entirely, however, is essentially turning away from three centuries of the development of mathematical analysis and its application to a multitude of problems of real practical importance, and should be aware of this.

3. <u>Jacobians and Newton's method</u>. The idea of the derivative of a functional as presented previously (see, in particular, (3.52) and (3.53) for conditions on derivatives), can be generalized in a natural way to functions (or <u>operators</u>) f which map a domain D in the real n-dimensional vector space R^n into R^m, concisely put, f: $D \subset R^n \to R^m$. In vector notation, for $x = (x_1,x_2,\ldots,x_n)^T$ in R^n, one has $f(x) = (f_1(x),f_2(x),\ldots,f_m(x))^T \in R^m$. The m×n matrix, formed by differentiation of the ith component of f with respect to the jth variable x_j to obtain its ijth element,

$$(6.24) \qquad f'(x) = \begin{bmatrix} \dfrac{\partial f_1}{\partial x_1} & \dfrac{\partial f_1}{\partial x_2} & \cdots & \dfrac{\partial f_1}{\partial x_n} \\ \cdots & \cdots & \cdots & \cdots \\ \dfrac{\partial f_m}{\partial x_1} & \dfrac{\partial f_m}{\partial x_2} & \cdots & \dfrac{\partial f_m}{\partial x_n} \end{bmatrix} = \begin{bmatrix} \dfrac{\partial f_i(x)}{\partial x_j} \end{bmatrix}$$

is called the <u>Jacobian matrix</u> of the transformation (or <u>operator</u>) f at x. As indicated, all the partial derivatives $\partial f_i/\partial x_j$, $i = 1,\ldots,m$; $j = 1,\ldots,n$, are to be evaluated at the point x. The construction of the Jacobian matrix requires the calculation of m·n partial derivatives; even for moderate values of m and n, this product is large enough to give an overwhelming advantage to automatic methods for differentiation of the m functions f_1, f_2, $\ldots$, f_m to obtain subroutines for evaluation of the required partial derivatives as compared to manual differentiation and coding, which also becomes increasingly susceptible to error as m and n grow larger. It was, in fact, the desire to compute Jacobian matrices automatically which motivated the writing of the original CODEX programs [30], [76].

Comparison of the Jacobian matrix (6.24) with (6.3) reveals that each row of the Jacobian matrix is simply the derivative of the corresponding function at x, that is, the ith row of (6.24) is $f_i'(x)$ as a row vector. Thus, the value of the transformed vector f(x) and the Jacobian matrix f'(x) may be obtained as the result of an evaluation and a gradient computation applied to f_1, f_2, $\ldots$, f_m in turn. This gives rise to the <u>Jacobian code list</u> for f, which is the sequence of gradient code lists,

$$
\begin{array}{l}
\text{Gradient code list for } f_1 \\
\text{Gradient code list for } f_2 \\
\cdots \quad \cdots \quad \cdots \quad \cdots \quad \cdots \quad \cdots \\
\text{Gradient code list for } f_m \ .
\end{array}
$$

(6.25)

Shorthand variables, of course, are handled the same way as indicated in §1.

As an m×n matrix, the Jacobian matrix (6.24) is a linear operator from R^n into R^m according to the "row-by-column" rule for matrix multiplication. That is, if A = (a_{ij}) is an m×n matrix, and $z = (z_1,z_2,\ldots,z_n)^T \in R^n$, then the components of the transformed vector

(6.26)
$$y = Az = (y_1,y_2,\ldots,y_m)^T \in R^m$$

are given by

(6.27)
$$y_i = \sum_{j=1}^{n} a_{ij}z_j, \quad i = 1,2,\ldots,m.$$

As discussed in Chapter 3 (see (3.52)-(3.53)), the existence of the partial derivatives $\partial f_i(x)/\partial x_j$ is not sufficient for the Jacobian matrix (6.24) to be a Fréchet or Gâteaux derivative of f as an operator from R^n (or $D \subset R^n$) into R^m [62], [71]. A condition under which the <u>differential</u>

(6.28)
$$df = f'(x)\,dx, \quad dx \in R^n,$$

approximates the <u>difference</u>

(6.29)
$$\Delta f = f(x + dx) - f(x)$$

as dx → 0 must also be satisfied. This condition is formulated in terms of a <u>norm</u> in in finite-dimensional vector spaces which takes the place of the absolute values in (3.52) and (3.53), which were used since the functional f considered there takes on values in $R = R^1$, the space of real numbers. The <u>maximum norm</u>,

(6.30)
$$\|x\| = \frac{max}{(i)} |x_i| ,$$

sometimes denoted by $\|x\|_\infty$, is probably the easiest norm to compute, and is satisfactory for most numerical applications. However, for arbitrary norms for R^m and R^n (in the numerator and denominator of the following expression, respectively),

(6.31)
$$\lim_{\|dx\| \to 0} \frac{\|f(x + dx) - f(x) - f'(x)dx\|}{\|dx\|} = 0$$

is the condition for the Jacobian matrix $f'(x)$ to be the <u>Fréchet derivative</u> of f at x, while for $dx = \tau h$, h an arbitrary unit vector in R^n ($\|h\| = 1$), the condition for $f'(x)$ to be the <u>Gâteaux derivative</u> of f at x is

(6.32)
$$\lim_{\tau \to 0} \|f(x + \tau h) - f(x) - f'(x)h\|/\tau = 0.$$

The idea of the approximation of the difference Δf by the differential df leads directly to the formulation of <u>Newton's method</u> for the solution of systems of nonlinear equations.

The basic idea is the following: Suppose that one wants to find a solution $x^* = (x_1^*, x_2^*, \ldots, x_n^*) \in R^n$ of the equation

(6.33)
$$f(x) = 0,$$

where $f: D \subset R^n \to R^n$; that is, (6.33) is a concise notation for a system of n (nonlinear) equations in n unknowns of the form (6.15). Given an <u>initial approximation</u> x^0 to x^*, one may use (6.31) after setting

(6.34)
$$x^* = x^0 + dx$$

to obtain

(6.35)
$$f(x^*) = f(x^0) + f'(x_0)dx + o(\|dx\|),$$

where, by (6.31), $o(\|dx\|)$ denotes a vector in R^n which is unknown, but has the property that

(6.36)
$$\lim_{\|dx\| \to 0} \{o(\|dx\|)/\|dx\|\} = 0 ,$$

which means that $o(\|dx\|)$ is very small if the error dx in the initial approximation x^0 to x^* is sufficiently small, that is, if x^0 is a "good" approximation to the unknown solution x^* of the system (6.33). (For the purposes of this discussion, it has been assumed that $f'(x)$ is the Fréchet derivative of f at x, see [71], Chapter IV, for a more complete description of Newton's method and its computational implementation.)

Going on the assumption that x^* exists, so that $f(x^*) = 0$, and that x^0 is a good approximation to x*, the system of equations (6.35) can be approximated by the <u>linear</u> system

(6.37)
$$f'(x^0) \delta x = - f(x^0) ,$$

where δx is an approximation to dx. This system is then solved, if possible, for

(6.38)
$$\delta x = [f'(x^0)]^{-1}(- f(x^0)) ,$$

after which

(6.39)
$$x^1 = x^0 + \delta x$$

is taken as the <u>improved approximation</u> to x*. From a computational standpoint, it should be noted that it is more efficient in most cases to solve the linear system (6.37) by some suitable method, such as Gaussian elimination with pivoting, rather than to perform the inversion of the matrix $f'(x^0)$ as indicated in (6.38) [23]. The inverse of the Jacobian matrix may be needed for theoretical purposes, however, or for certain interval computation schemes which will be discussed in the following chapters. For the present, (6.38) will be regarded as symbolic notation for the solution of (6.37). After x^1 has been calculated, then it may be taken as the initial approximation to x*, and the above process repeated to obtain another improved approximation x^2 to x*, provided that the construction and solution of the linear system corresponding to (6.37) can be carried out successfully. In the notation for vectors used here, the superscripts 0, 1, 2, ... are used simply for enumeration, and do not imply exponentiation; in general, x^k will be used to denote the vector

(6.40)
$$x^k = (x_1^{(k)}, x_2^{(k)}, \ldots, x_n^{(k)})^T, \quad k = 0, 1, 2, \ldots .$$

The process of generating the sequence of vectors $\{x^k\}$ by the method described above, that is, constructing and solving the linear equations

(6.41)
$$f'(x^k) \delta x^k = - f(x^k)$$

and then taking

(6.42)
$$x^{k+1} = x^k + \delta x^k, \quad k = 0, 1, 2, \ldots,$$

is called <u>Newton's method</u>, and the corresponding sequence of vectors $\{x^k\}$ is called the <u>Newton sequence</u>. Of course, in an actual computation, only a finite number of terms of the Newton sequence can be computed. One feature of Newton's method which recommends it in practice is that it converges <u>quadratically</u>, that is, under fairly mild conditions, a constant M exists such that

(6.43)
$$\| x^* - x^{k+1} \| \leq M \cdot \| x^* - x^k \|^2 ,$$

so that it may be possible to achieve a high accuracy in just a few iterations. This is especially important in spaces of moderate to high dimension n, in which the calculations involved to evaluate the Jacobian matrix $f'(x^k)$ and solve the linear system (6.41) may be of considerable expense. Conditions under which the quadratic rate of convergence of Newton's method is assured are given in [61] and [71], among other places, and a set of optimal error bounds are given in [24]. It is also worthy of note that the conditions for the convergence of the Newton sequence also guarantee the existence of x*, and may be computationally verified at the initial approximation x^0, as will be described later in connection with the program NEWTON. This latter computation, which amounts to the rigorous verification of the hypotheses of a mathematical theorem on the computer, also involves the calculation of derivatives automatically.

It is fair to say that if manual methods are used for differentiation, then the usefulness of Newton's method is limited to small or special systems of nonlinear equations. Software for automatic differentiation extends the usefulness of this powerful method and its theory to a much wider range of problems. Alteration of the computational method by approximation of the Jacobian matrix (for example, by the use of finite differences instead of derivatives) is, of course, an alternative to the use of derivatives. The resulting procedures are frequently called quasi-Newton methods [20]; their use may require sacrifice of the rate of convergence or the theoretical simplicity of the exact method with regard to guarantee of existence of solutions or ease in obtaining error bounds. However, such approximate methods have been found to be effective in practice [21]; at the present time, there are not enough valid comparsions of quasi-Newton methods with Newton's method using automatically computed derivatives to draw any firm conclusions.

4. Second derivatives: Hessian matrices and operators. Returning to the optimization problems discussed in §2, suppose that it is desired to apply Newton's method to the solution of the system of equations (6.15). (Much the same considerations as given below apply to the system (6.17)(b)-(6.18) obtained for constrained optimization problems; for simplicity, only the unconstrained case will be discussed in detail.) The Jacobian matrix of the system (6.15) has elements

(6.44)
$$f_{i,j}(x_1,x_2,\ldots,x_n) = \frac{\partial^2 f(x)}{\partial x_i \partial x_j} \, , \quad i,j = 1,2,\ldots,n.$$

In (6.44), the convention

(6.45)
$$\frac{\partial^2 f}{\partial x_i \partial x_j} = \frac{\partial}{\partial x_j}\left(\frac{\partial f}{\partial x_i}\right)$$

has been adopted, as in [71], Chapter 3, which also contains the essential background material for this discussion.

The matrix with components (6.44) is called the Hessian matrix of the functional f at x, and will be denoted by f"(x). If the condition (6.31) is satisfied, then the Hessian matrix of f at x is a representation of the second Fréchet derivative of f at x. Under this assumption, the computation of the Hessian matrix of a function can be simplified by noting that the Hessian matrix f"(x) is symmetric [71], that is,

(6.46)
$$\frac{\partial^2 f}{\partial x_i \partial x_j} = \frac{\partial^2 f}{\partial x_j \partial x_i} \ , \quad i,j = 1,2,\ldots,n.$$

This means that the matrix

(6.47)
$$f''(x) = \left[\frac{\partial^2 f}{\partial x_i \partial x_j} \right]$$

consists of only $\frac{1}{2}n(n+1)$ distinct elements, rather than n^2, as in the case of a general Jacobian matrix. For $F = F(X1,X2,\ldots,Xn)$, let

(6.48)
$$\text{D2FDXiDXj} = \partial^2 F/\partial x_i \partial x_j, \quad i,j = 1,2,\ldots,n;$$

then the necessary distinct partial derivatives to form the Hessian matrix of f at x can be obtained from the sequence of code lists,

(6.49)

> Derivative code list for D2FDX1DX1
> Derivative code list for D2FDX1DX2
> ...   .
> Derivative code list for D2FDX1DXn
> Derivative code list for D2FDX2DX2
> ...   .
> Derivative code list for D2FDX2DXn
> ...   .
> Derivative code list for D2FDX(n-1)DX(n-1)
> Derivative code list for D2FDX(n-1)DXn
> Derivative code list for D2FDXnDXn ,

assuming, of course, that F and its gradient (or derivative) have already been processed into code lists. The list (6.49) will be called, of course, a <u>Hessian</u> <u>code</u> <u>list</u> for F, in <u>upper</u> <u>triangular</u> <u>form</u>, the latter designation referring to the distinct portion of the matrix actually computed. A Hessian code list for F in <u>lower</u> <u>triangular</u> form would consist of the sequence of code lists

(6.50)

> Derivative code list for D2FDX1DX1
> Derivative code list for D2FDX2DX1
> Derivative code list for D2FDX2DX2
> ...   .
> Derivative code list for D2DX(n-1)DX1
> ...   .
> Derivative code list for D2DX(n-1)DX(n-1)
> Derivative code list for D2FDXnDX1
> ...   .
> Derivative code list for D2FDXnDXn .

From the standpoint of computing the distinct elements of $f''(x)$, (6.49) and (6.50) are equivalent; it is important, of course, to be specific about the actual order in which the components of the Hessian matrix are computed and stored in any given application. The convention (6.45) has been adopted in the naming of the second derivatives in the above discussion; for example, the derivative code list for D2FDX2DX3 is obtained by differentiation of the derivative code list for DFDX2 (= $\partial F/\partial X2$) with respect to X3, and so forth.

As explained in [71], the second derivative of an operator $f: D \subset R^n \rightarrow R^m$ is actually a <u>bilinear operator</u> from R^n into R^m, which is a linear operator from R^n into the space of linear operators from R^n into R^m (the space of m×n matrices). The way to represent bilinear operators which is convenient for computational purposes is as an m×n×n array (a kind of "three-dimensional matrix"). Thus, if B is a bilinear operator from R^n into R^m, it can be written

$$(6.51) \qquad B = (b_{ijk}), \quad i = 1,2,\ldots,m; \; j,k = 1,2,\ldots,n.$$

The rule for transformation of a vector $x = (x_1,x_2,\ldots,x_n)^T \in R^n$ into an m×n matrix $A = Bx$ is

$$(6.52) \qquad a_{ij} = (Bx)_{ij} = \sum_{k=1}^{n} b_{ijk} x_k, \quad i = 1,2,\ldots,m; \; j = 1,2,\ldots,n.$$

The matrix $A = Bx = ((Bx)_{ij})$ can then transform an arbitrary vector $z \in R^n$ into a vector $y \in R^m$ with components

$$(6.53) \qquad y_i = \sum_{j=1}^{n} \sum_{k=1}^{n} b_{ijk} x_k z_j, \quad i = 1,2,\ldots,m.$$

This transformation is symbolized by

$$(6.54) \qquad y = (Bx)z = Bxz ,$$

and if $Bxz = Bzx$ for all $x,z \in R^n$, then the bilinear operator B is said to be <u>symmetric</u> [71].

Symmetric bilinear operators appear in analysis as second Fréchet derivatives of operators $f: D \subset R^n \rightarrow R^m$, as indicated above. The second derivative $f''(x)$ of f at x has the representation

$$(6.55) \qquad f''(x) = \left[\frac{\partial^2 f_i(x)}{\partial x_j \partial x_k} \right], \quad i = 1,2,\ldots,m; \; j,k = 1,2,\ldots,n,$$

and, in addition, has to satisfy the relationship

$$(6.56) \qquad \lim_{\|dx\| \to 0} \frac{\|f'(x + dx) - f'(x) - f''(x)dx\|}{\|dx\|} = 0$$

in terms of the first derivative f' of f for $dx \in R^n$. The norm in the numerator of (6.56) is a norm for the space of linear operators from R^n into R^m [71], that is, a norm for m×n matrices. The particular norm chosen must be <u>consistent</u> with the norms for R^n and R^m in the sense that if $A = (a_{ij})$ is an m×n matrix, and $y = Ax$ is the vector in R^m obtained by operating with A on the vector $x \in R^n$, then

(6.57)
$$\|Ax\| \le \|A\| \cdot \|x\|$$

where each norm is taken in the space to which the item enclosed belongs. For example, if the maximum norm (sometimes also called the uniform norm) is taken in R^m and R^n, then a consistent norm for m×n matrices $A = (a_{ij})$ is

(6.58)
$$\|A\| = \frac{\max}{(i)} \sum_{j=1}^{n} |a_{ij}| \; .$$

As is the case with the vector norm (6.30), the matrix norm (6.58) is one of the easiest to compute; furthermore, it is a precise bound in the sense that equality is attained in (6.57) for some unit vector $\hat{x}$. To see this, suppose that (6.58) holds for the index $i = i_0$, and take

(6.59)
$$\hat{x}_j = \text{sign}(a_{i_0 j}) = \begin{cases} -1 \text{ if } a_{i_0 j} < 0, \\ 0 \text{ if } a_{i_0 j} = 0, \\ 1 \text{ if } a_{i_0 j} > 0, \end{cases}$$

$j = 1,2,\ldots,n$. The vector $\hat{x} = (\hat{x}_1, \hat{x}_2, \ldots, \hat{x}_n)$ obtained in this way is then a unit vector with the desired property.

The Hessian matrix of a functional f is the special case of the second derivative (6.55) obtained for $m = 1$. The property of symmetry also holds for larger values of m, since

(6.60)
$$\frac{\partial^2 f_i(x)}{\partial x_j \partial x_k} = \frac{\partial^2 f_i(x)}{\partial x_k \partial x_j} \; , \quad i = 1,2,\ldots,m; \; j,k = 1,2,\ldots,n,$$

provided that (6.56) holds [71]. Thus, the second Fréchet derivative of a twice differentiable operator f will be a symmetric bilinear operator. Consequently, the array (6.55) will contain less than $m \cdot n^2$ distinct elements; as derived above for the case of Hessian matrices, there will actually be $\frac{m}{2} \cdot n(n+1)$ distinct second partial derivatives which make up the array representing f"(x), which will be called the Hessian operator of f at x. A code list for the Hessian operator of the operator $f = (f_1, f_2, \ldots, f_m)^T$ from R^n into R^m can be constructed by using the Hessian code lists for the individual components f_i of f, $i = 1,2,\ldots,m$, considered as functionals:

(6.61)

Hessian code list for f_1

Hessian code list for f_2

$\cdots \quad \cdots \quad \cdots \quad \cdots \quad \cdots$

Hessian code list for f_m .

Since the actual computation of the Hessian operator of f requires the function and gradient code lists for the components f_1, f_2, $\ldots$, f_m of f in addition to (6.61), the total amount of code produced (in terms of the number of lines in the code lists, and the number of intermediate variables required for labels) can be extensive, perhaps even overwhelming. Higher derivatives than second, which are multilinear operators represented by arrays of even larger dimensions [71], present even more of a challenge. However, if needed, code lists for these operators can be formed from the function code lists for f_1, f_2, ..., f_m by the methods discussed above. All higher derivatives are also symmetric multilinear operators [71], a fact which can be used to reduce the complexity of their calculation by a certain extent.

If the problem at hand requires only values of $f(x)$, $f'(x)$, and $f''(x)$ for some given operator $f\colon D \subset R^n \to R^m$ (usually, $m = 1$ or $m = n$), then the code lists

(6.62)
$$\text{Function code list for } f$$
$$\text{Jacobian code list for } f$$
$$\text{Hessian code list for } f \text{ ,}$$

provide the necessary means to obtain the required quantities. As pointed out above, the lists (6.62) may actually be of considerable length. The original program NEWTON [25], [26], see also [71], Chapter 4, used this approach; the Jacobian matrix was used in the iterative solution of a given system of nonlinear equations of the form (6.15) by Newton's method, as outlined above. A description of the use of the Hessian operator in connection with verification of existence of solutions of the system and estimation of the error of approximate solutions computed will be given in a later chapter.

The code for obtaining $f(x)$, $f'(x)$, $f''(x)$ (if only these values are wanted) can be made more compact, once the code list for the function f has been produced, by using ideas which are based on the concept of differential code lists, as introduced in §1 of Chapter 4. In order to obtain both first and second derivatives, second differentials, which will be explained below, are used in addition to the first differentials described previously. These ideas are implemented in the version of the program NEWTON described by Kuba and Rall [39]. Another observation which should be made at this point about code lists for first and second derivatives of functions of several variables is that while most, if not all, lines in a code list for a function $f(x_1)$ of a single variable will contain entries which depend on x_1, it is to be expected that many of the lines in a code list for a function $f = f(x_1, x_2, \ldots, x_n)$ of several variables will depend on variables other than, say, x_1. In forming the code lists for the first and second derivatives of f with respect to x_1, the other variables and lines in the code list dependent on only them will be treated as constants, so these lines will not produce corresponding lines in the (packed) derivative code list for $\partial f/\partial x_1$. Thus, if "on the average", about $1/n$ of the lines in the function code list for f depend on any single variable x_i, $i = 1, 2, \ldots, n$, then about k/n lines will appear in the code list for $\partial f/\partial x_i$, where k is the average "expansion factor" in going from function code lists to derivative code lists for functions of a single variable (one expects that $1 < k < 2$). Therefore, the gradient code list for f will have only about k times the number of lines in the function code list under the above assumption, which is the same factor encountered in going from the function code list to the derivative code list for functions of a single variable.

Returning to the introduction of second differentials, these may be defined in terms of second (Fréchet) derivatives in the same way that differentials are related to first derivatives. Suppose, for simplicity, that $f = f(x,y)$ is a function of two variables only. (This is actually the most general case needed, since the entry in each line of a code list depends on at most two labels, and the chain rule applies to their differentials.) The second derivative of f, written according to the notation in [71] to produce a row vector (linear functional) as the result of operation on a column vector, is at $z = (x,y)^T \in R^2$,

$$(6.63) \qquad f''(z) = \left(\frac{\partial^2 f}{\partial x^2} \quad \frac{\partial^2 f}{\partial y \partial x} \ \middle|\ \frac{\partial^2 f}{\partial x \partial y} \quad \frac{\partial^2 f}{\partial y^2} \right) .$$

Operation on an (arbitrary) differential vector $dz = \binom{dx}{dy}$ with (6.63) gives the row vector

$$(6.64) \qquad f''(z) \binom{dx}{dy} = \left(\frac{\partial^2 f}{\partial x^2} \cdot dx + \frac{\partial^2 f}{\partial y \partial x} \cdot dy \quad \frac{\partial^2 f}{\partial x \partial y} \cdot dx + \frac{\partial^2 f}{\partial y^2} \cdot dy \right) .$$

Operation on a second differential vector, say $\binom{dxx}{dyy}$, gives the second (Fréchet) differential of f,

$$(6.65) \qquad d^2 f = \frac{\partial^2 f}{\partial x^2} \cdot dx \cdot dxx + \frac{\partial^2 f}{\partial y \partial x} \cdot dy \cdot dxx + \frac{\partial^2 f}{\partial x \partial y} \cdot dx \cdot dyy + \frac{\partial^2 f}{\partial y^2} \cdot dy \cdot dyy ,$$

in terms of the differential vectors $\binom{dx}{dy}$ and $\binom{dxx}{dyy}$. Given a code list for $d^2 f$, the various second partial derivatives of f can be obtained by selection of the differential vectors; for example, with $\binom{dx}{dy} = \binom{1}{0}$, $\binom{dxx}{dyy} = \binom{0}{1}$, the corresponding value of $d^2 f$ is

$$(6.66) \qquad d^2 f = f''(z) \binom{1}{0} \binom{0}{1} = \frac{\partial^2 f}{\partial y \partial x} .$$

The other second partial derivatives of f can be picked out of the expression (6.65) in the same way.

Another way to obtain the second differential of f is as the differential of its first differential df. One has, from (3.43),

$$(6.67) \qquad d(df) = \frac{\partial(df)}{\partial x} \cdot dxx + \frac{\partial(df)}{\partial y} \cdot dyy ,$$

and, for example,

$$(6.68) \qquad \frac{\partial}{\partial x}(df) = \frac{\partial^2 f}{\partial x^2} \, dx + \frac{\partial f}{\partial x} \cdot \frac{\partial}{\partial x}(dx) + \frac{\partial^2 f}{\partial x \partial y} \cdot dy + \frac{\partial f}{\partial y} \cdot \frac{\partial}{\partial x}(dy),$$

and, since dx and dy are assumed to be independent of x (and y), one has

(6.69)
$$\frac{\partial}{\partial x}(df) = d_x(df) = \frac{\partial^2 f}{\partial x^2} \cdot dx + \frac{\partial^2 f}{\partial x \partial y} \cdot dy \ ,$$

and (6.65) becomes

(6.70)
$$d^2 f = d(df) = d_x(df) \cdot dxx + d_y(df) \cdot dyy \ ,$$

where the symmetry of the mixed second partial derivatives has been used. The second partial derivatives of f with respect to x may be obtained from (6.69) by setting dx = 1, dy = 0, and then dx = 0, dy = 1 in turn, as before. Suppose now that the gradient code list for f has been formed from the differential code list for df. Then, the code list for (6.69) is simply the differential code list for the function $\partial f / \partial x$; that is, $d(\partial f / \partial x) = \partial(df) / \partial x$.

It is possible to compress the coding involved in the computation of second derivatives considerably by using subroutines or producing code lists based on explicit formulas for the first and second derivatives of arithmetic operations and library functions. For example, for multiplication, one would have the sequence of formulas

(6.71)
```
              T = U*V
        DXT = U*DXV + V*DXU,  DYT = U*DYV + V*DYU
        DYDXT = U*DYDXV + DXV*DYU + V*DYDXU + DXU*DYV
```

to evaluate corresponding to the line T = U*V in the code list in order to calculate the second partial derivative $\partial^2 T / \partial Y \partial X = \partial^2(U*V) / \partial Y \partial X$. As well as the current values of X and Y, the values of the derivatives DXX = 1, DYX = 0, DXY = 0, DYY = 1 are to be supplied. (These are also differentials, DXX = $(\frac{\partial X}{\partial X}) \cdot$DXX, so it is a matter of taste as to whether the result of producing a code list based on formulas of the type (6.71) is called a differential or a derivative code list. It is termed a differential list here because the pattern of assigning values to DXX, etc. follows the one discussed in connection with the second differential (6.65).)

If the gradient code list of F has been formed, then the Hessian code list requires only code based on the third line of (6.71), that is, code lists in which the sublist, for example

(6.72)
```
              T1 = U*DYDXV
              T2 = DXV*DYU
              T3 = T1 + T2
              T4 = V*DYDXU
              T5 = T3 + T4
              T6 = DXU*DYV
          DYDXT = T5 + T6
```

corresponds to the line T = U*V in the function code list

The variables X, Y cited above are arbitrary; if F = F(X1,X2,...,XN), then the (upper triangular) Hessian code list can be obtained from (6.72) by taking

$$
\begin{array}{ll}
X = X1; & Y = X1, X2, \ldots, XN; \\
X = X2; & Y = X2, X3, \ldots, XN; \\
\ldots \quad \ldots & \ldots \quad \ldots \quad \ldots \quad \ldots \quad \ldots \\
X = X(N-1); & Y = X(N-1), XN; \\
X = XN; & Y = XN;
\end{array}
$$

(6.73)

in turn in (6.72). From this, it follows that to obtain the code list for the second derivative $\partial^2 F/\partial X1 \partial X3$, one would take

(6.74)
$$
\begin{array}{l}
DXX1 = DXX3 = 1, \\
DXX2 = DXX4 = DXX5 = \ldots = DXXN = 0 ,
\end{array}
$$

while for $D2FDX2DX2 = \partial^2 F/\partial X2^2$, the corresponding substitution would be

(6.75)
$$
\begin{array}{l}
DXX2 = 1, \\
DXX1 = DXX3 = DXX4 = \ldots = DXXN = 0 .
\end{array}
$$

Another possibility, which is the option actually selected in the program [39], is to code the formulas of the type (6.71) as subroutines such as

(6.76)
$$
MULT(X,Y,U,V,T,DXT,DYT,DYDXT) ,
$$

and use the function code list to construct a sequence of calls to these subroutines, as in the case of the generation of Taylor coefficients. In (6.76), the subroutine should be written in such a way as to produce DYT = DXT if entered with Y = X, without unnecessary computation. It is only necessary to have one formula in the second line of (6.71); if X = X1, Y = X3, for example, then these values may be used successively in the single formula

(6.77)
$$
DXT = U*DXV + V*DXU ,
$$

and, of course, if X = Y = X2, say, then (6.77) would only be computed once.

Formulas corresponding to (6.71) for gradient and Hessian computation (the gradient computation simply omits the third line) are given in Table VI.1, as actually used in the program [39]. An examination of the amount of computation performed by use of the subroutines in the evaluation of the Hessian matrix of a function F of N variables X1, X2, ..., XN leads to the following conclusions:

1°. The value of $F(X1,X2,\ldots,XN)$ is calculated $\frac{N}{2} \cdot (N+1)$ times, that is, once for each distinct element of the Hessian matrix of F.

2°. The gradient of F, $\nabla F = (DX1F, DX2F, \ldots, DXNF)^T$, is calculated N times, in the following way: To calculate the row of the Hessian matrix corresponding to X1, that is, the elements corresponding to the first row of (6.73), one computes DX1F a total of N times; in the second row, corresponding to X2, the derivative DX2F is obtained N-1 times, however, it was computed once in the first row to get DX1DX2F, so the number of times DX2F is evaluated is also N, and so on.

If storage is available, then a single evaluation of the function code list and storage of the results would eliminate the duplication encountered in 1°; similarly,

a single evaluation of the gradient code list and storage of the results would eliminate the duplication in both 1° and 2°, and subroutines based on the third line of
(6.71) could be called to compute the elements of the Hessian matrix of F. Thus,
compact coding of the Hessian computation costs time. (See, however, the method suggested in the next section.) An alternative is to use more storage; the decision as
how to make this trade-off will depend on the characteristics of the system being
used.

The following table gives formulas for the first and second differentials corresponding to code list entries involving arithmetic operations, library functions,
and constants (denoted by C). The collection of recipes given in Table VI.1 can be
used to cook up code lists or subroutines to be called for gradient or Hessian computations, or both.

TABLE VI.1. FIRST AND SECOND DIFFERENTIALS

Entry in Code List	Formulas for First and Second Differentials
1. T = C	DXT = 0,
	DYDXT = 0;
2. T = C + U	DXT = DXU,
	DYDXT = DYDXU;
3. T = C*U	DXT = C*DXU,
	DYDXT = C*DYDXU;
4. T = U + V	DXT = DXU + DXV,
	DYDXT = DYDXU + DYDXV;
5. T = U - V	DXT = DXU - DXV,
	DYDXT = DYDXU - DYDXV;
6. T = U*V	DXT = DXU*V + U*DXV,
	DYDXT = DYDXU*V + DXU*DYV + DYU*DXV +
	+ U*DYDXV;
7. T = U/V	DXT = (DXU - T*DXV)/V,
	DYDXT = (DYDXU - DXT*DYV - DYT*DXV -
	- T*DYDXV)/V;
8. T = U**C	DXT = C*DXU*U**(C-1),
	DYDXT = C*(C-1)*DYU*DXU*U**(C-2) +
	+ C*DYDXU*U**(C-1);
9. T = EXP(U)	DXT = DXU*T
	DYDXT = DYDXU*T + DXU*DXT;
10. T = SIN U	DXT = DXU*Z
Z = COS(U)	DXZ = - DXU*T
	DYDXT = - DYU*DXU*T + DYDXU*Z
	DYDXZ = - DYU*DXU*Z - DYDXU*T;

TABLE VI.1.   (CONTINUED)

Entry in Code List	Formulas for First and Second Differentials
11. T = LOG(U)	DXT = DXU/U,
	DYDXT = (DYDXU - DYU*DXT)/U;
12. T = SQRT(U)	DXT = (DXU/2)*T/U,
	DYDXT = ((DYDXU/2)*T + (DXU/2)*DYT -
	- DYU*DXT)/U;
13. T = TAN(U)	DXT = DXU*(1 + T*T),
	DYDXT = DYDXU * (1 + T*T) + 2*DXU*DYT*T;
14. T = ATAN(U)	DXT = DXU/(1 + U*U),
	DYDXT = (DYDXU - 2*U*DYU*DXT)/(1 + U*U).

5.   Projects for Chapter 6.

1°. In your favorite language, write software to produce gradient and Hessian code lists from the function code list for a functional f.

2°. Write software to produce the Jacobian and the Hessian code lists for an operator f = $(f_1, f_2, \ldots, f_m)$ from R^n into R^m.

3°. Write software to produce the system of equations (6.15) for unconstrained optimization, given a formula for the functional f.

4°. Write software to produce the system of equations (6.17)(b)-(6.18) for constrained optimization, given formulas for f and the constraints g_1, g_2, ..., g_m. The result should produce the system (6.15) in case the set of constraints is empty.

5°. Code the formulas in Table VI.1 for gradient, Jacobian, and Hessian computations.

6°. Code the formulas in Table VI.1 to produce the first and second differentials df and d^2f of a functional defined by a formula.

7°. Code more efficient gradient and Hessian calculations which compute each quantity only once. For example, for

(6.78)   T = U*V,

the gradient would result from setting X = Xi, i = 1,2,...,N in turn, and use of

(6.79)   DXT = U*DXV + V*DXU ;

similarly, the Hessian could be obtained by setting X = Xi, i = 1,2,...,N, and Y = Xj, j = i, i+1, ..., N, and using

(6.80)   DYDXT = U*DYDXV + DXV*DYU + V*DYDXU + DXU*DYV,

based on (6.71). Code this procedure, and compare the speed with the straightforward method used in the program [39].

8°. Adapt the procedure in 7° to Jacobian and Hessian computations for operators f from R^n into R^m.

CHAPTER VII

AUTOMATIC ERROR ANALYSIS

1. <u>Errors in computation</u>. In actual numerical computation using digital computers, the user is bedeviled by errors which arise in various fashions, including the
following:

1°. <u>Errors in data</u>. In the context of formula evaluation and differentiation,
this means that the values of the constants, parameters, and variables may not be
known or represented exactly, so that one has to compute with approximations to them.
The error coming from inaccuracy in the data is called <u>data error</u>.

2°. <u>Roundoff error</u>. Computer arithmetic is not carried out exactly, even on
exactly known arguments, so that most calculations of the results of arithmetic operations and library functions will be inaccurate, since all results are required to
be expressed in terms of numbers with a prescribed format, such as a fixed number of
"significant" digits. The actual computation thus results in an approximation to the
true result of the transformation or operation being performed, and the difference
between the result obtained and the true result is called the <u>roundoff error</u> of the
computation.

3°. <u>Truncation error</u>. Many mathematical quantities, such as integrals, derivatives, and various algebraic and transcendental functions, are actually defined only
as the limits of infinite sequences of operations. Unless, as in the case of differentiation of simple functions, rules are available which give the values of these limits explicitly, their computation can only be carried out to a finite point of the
sequence defining their values. The resulting error of the approximate result obtained by actual computation is called the <u>truncation error</u>.

Ordinarily, a result produced by execution of a computer program will be contam-
inated by one or more of the above types of error. Furthermore, it is not possible
to draw a clear line of distinction between these types of error, or eliminate them
completely. For example, numbers such as 1/3 or π which appear in a formula may be
impossible to represent exactly as numbers in the system used by a given type of computer. The resulting error could be considered to be due to data, roundoff, or truncation error. Similarly, the error in computation of sin(x) by a library subroutine
(hopefully minimal), can be attributed to either roundoff error, or truncation error
in the computation of the transcendental sine function. Whatever the source, error
is pervasive in digital computation [66], [67], and the accuracy of the result of hun-
dreds or thousands of inexact computations on inaccurate data is always questionable
without some investigation. The problem of the analysis of error of numerical results

is one which is fundamental to statements about the reliability of the computation (except for those with blind faith in the output of computers). Unfortunately, this problem is among the most laborious to solve in numerical analysis, and is considered by a number of mathematicians to be deadly dull as compared to the exciting field of design of algorithms to solve new and important problems. Consequently, error analysis is a prime candidate for automation to avoid the drudgery involved; several methods for automatic error analysis will be considered in this chapter.

2. Interval arithmetic. It turns out that one method which presents the possibility to carry out the error analysis automatically in connection with the computation itself is the use of interval arithmetic, as developed by R. E. Moore (see, for example, [51], [53], [55]). This is a large (and growing) field, but the basic idea for computational purposes is the following: Given a function f(x) of a single real variable to start with, suppose it is known that x belongs to an interval X = [a,b], that is, $a \leq x \leq b$, where a and b are machine numbers, which means that they can be represented exactly in whatever system of numeration is being used in the actual computation. The range of the function f on the interval X is of course the set

$$(7.1) \qquad f(X) = \{y \mid y = f(x), a \leq x \leq b\} ,$$

which in general may not be known exactly nor be representable in terms of machine numbers. However, if it is always possible to find machine numbers c, d, such that

$$(7.2) \qquad c \leq f(x) \leq d, \quad a \leq x \leq b,$$

then the set (7.1) is contained in the interval Y = [c,d], and one may define an interval function F associated with f by the transformation of the interval [a,b] into the interval [c,d] in the above fashion, that is,

$$(7.3) \qquad Y = F(X) .$$

This interval function contains the real function f in the sense that $f(X) \subset F(X)$. It will also be required that F be inclusion monotone, that is, if the intervals X, Z are such that $X \subset Z$, then $F(X) \subset F(Z)$. An inclusion monotone interval function which contains a real function f is called an interval extension of f. Of course, in practice one would want F to be an "accurate" extension of f in the sense that the interval F(X) containing f(X) is "as small as possible". The importance of interval analysis in this fashion for error estimation is that if one can compute an interval extension of a given real function, then $f(x) \in F(X)$ if $x \in X$, and one can then derive bounds for the error in the computation of f(x) from the interval results using the width w(F(X)) of the interval F(X), that is,

$$(7.4) \qquad w(F(X)) = w([c,d]) = d - c .$$

An interval arithmetic for a given computer consists of interval extensions of arithmetic operations and a selected set of library functions. An interval arithmetic package would then consist of the subroutines for input and output, and other housekeeping chores. Such interval arithmetic packages have been developed at the

Mathematics Research Center since 1965 by Reiter [77], [79], Yohe, and others [4], [92], [93], [94], and recent versions can be used with the AUGMENT precompiler [12], [13], [14], [15], [16], to allow the user to declare quantities to be of TYPE INTERVAL in FORTRAN programs. One important feature of interval arithmetic in general is di-rected rounding; results of operations are rounded up to the smallest machine number larger than the result, and down to the largest machine number less than the result to obtain a machine interval of minimal width in which the result is guaranteed to lie. For example, if the difference d - c in (7.4) turns out not to be an exact machine number, then it would be increased by directed rounding to the smallest ma-chine number μ such that $\mu \geq w(F(X))$. See Table VII.1 for a list of basic operations.

TABLE VII.1. FUNDAMENTAL INTERVAL OPERATIONS

Basic Operations   Interval Result (See Note 7.1)

(7.5) $[a,b] + [c,d]$ $=$ $[a + c , b + d];$

(7.6) $k*[a,b]$ $=$ $[k*a , k*b]$ if $k \geq 0,$

                            $=$ $[k*b , k*a]$ if $k < 0;$

(7.7) $[a,b] - [c,d]$ $=$ $[a - d , b - c];$

(7.8) $[a,b]*[c,d]$ $=$ $[\min\{a*c,a*d,b*c,b*d\} , \max\{a*c,a*d,b*c,b*d\}];$

(7.9) $[a,b]^{-1}$ $=$ $[b^{-1} , a^{-1}]$ if $ab > 0,$

                                     undefined if $ab \leq 0;$

(7.10) $[a,b]^2$ $=$ $[\min\{a^2,b^2\} , \max\{a^2,b^2\}]$ if $ab > 0,$

                            $=$ $[0 , \max\{a^2,b^2\}]$ if $ab \leq 0;$

Compound Operation

(7.11) $[a,b]/[c,d]$ $=$ $[a,b]*[c,d]^{-1}$ (undefined if $cd \leq 0$).

Note 7.1. If the indicated endpoint of an interval operation is not a machine number, then directed rounding is used to produce the smallest interval $[\alpha,\beta]$ with machine number endpoints which contains the result. It is assumed that a,b,c,d, and k are machine numbers. A machine number k may be identified with the degenerate ma-chine interval $[k,k]$, and one writes $k = [k,k]$.

Note 7.2. From (7.6) and (7.7), one can write

(7.12) $[a,b] - [c,d]$ $=$ $[a,b] + (-1)*[c,d];$

using (7.8), (7.6) can be written as

(7.13) $k*[a,b]$ $=$ $[k,k]*[a,b]$.

Note 7.3. The definition of the square of an interval given in (7.10) is not, in general, equal to $[a,b]*[a,b]$; however, it does provide an interval extension of the real function defined by $f(x) = x^2$. The interval operation $[a,b]^2$ defined by (7.10) can be used to obtain $[a,b]^{**}I$ for I an integer by the method discussed in Chapter 4, §4 (see equations (4.53)-(4.70)). The result will be an interval extension of the real function defined by $f(x) = x^I$, I an integer.

The code list representation of functions is suitable for the use of interval arithmetic as well as for differentiation; if the code list is executed in interval arithmetic, then the resulting interval will contain the result of execution of the code list in _exact_ real arithmetic. This of course, assumes that interval subroutines are available for all arithmetic operations and library functions allowed in the code list. In fact, the theory of interval analysis [53], [55], allows one to make the following assertion.

The Fundamental Theorem of Automatic Interval Error Analysis (R. E. Moore): If all operations in a code list for the real function f are replaced by the corresponding interval operations, then the result is a code list for an interval extension F of f.

Thus, the execution of code lists for derivatives and Taylor coefficients of a given function in interval arithmetic will provide intervals which contain the exact values of the corresponding quantities for each value of x in the input interval X.

The above comments and assertions generalize immediately to functions of several variables. The _magnitude_ $|X|$ of an interval $X = [a,b]$ is defined to be

$$(7.14) \qquad |X| = |[a,b]| = \max\{|a|, |b|\}.$$

Thus, for an _interval vector_ $X = (X_1, X_2, \ldots, X_n)$, one may define the _norm_

$$(7.15) \qquad \|X\| = \max_{(i)}\{|X_i|\},$$

as in (6.30). Similarly, for _interval matrices_ $A = (A_{ij})$, one has

$$(7.16) \qquad \|A\| = \max_{(i)}\{\sum_{j=1}^{n}|A_{ij}|\},$$

which is analogous to (6.58), and for which $\|AX\| \le \|A\| \cdot \|X\|$ holds [53], [55]. Also, for _interval bilinear operators_ $B = (B_{ijk})$ [39], [71], one has for

$$(7.17) \qquad \|B\| = \max_{(i)}\{\sum_{j=1}^{n}\sum_{k=1}^{n}|B_{ijk}|\},$$

that for $A = BX$ with $\|A\|$ defined by (7.16) and $\|X\|$ by (7.15),

$$(7.18) \qquad \|A\| = \|BX\| \le \|B\| \cdot \|X\|,$$

and thus for $Y = BXZ$, $Z = (Z_1, Z_2, \ldots, Z_n)$,

$$(7.19) \qquad \|Y\| = \|BXZ\| \le \|B\| \cdot \|X\| \cdot \|Z\|,$$

an expression of the <u>consistency</u> of the norms (7.15), (7.16), and (7.17) [39], [71],
as in the case of ordinary vector, matrix, and operator norms.

Use of interval methods of computation can thus take much of the drudgery out of
error estimation if properly conceived and executed, since one is provided with an
interval computed by inexact operations on inaccurate data which contains the real
number (or vector, matrix, etc.) that would have been obtained if the operations on
exact data were performed with infinite precision. Given such an interval as output,
one can take the <u>midpoint</u> (or <u>arithmetic point</u>, or <u>arithmetic mean</u>)

$$(7.20) \qquad x_A = m([a,b]) = \frac{1}{2}(a + b)$$

as an approximation to a proported exact value $x \in [a,b]$. The <u>absolute error</u> of x_A
as an approximation to x is then bounded by

$$(7.21) \qquad |x - x_A| \le \frac{1}{2}(b - a) = \frac{1}{2} \cdot w([a,b]),$$

the so-called <u>half-width</u> of the interval [a,b]. An alternative form of representation of the interval [a,b] is in the form of the expression

$$(7.22) \qquad [a,b] = \frac{1}{2}(a + b) \pm \frac{1}{2}(b - a) = m([a,b]) \pm \frac{1}{2} \cdot w([a,b]).$$

The program [39] for Newton's method and the program [28] for numerical integration
by interval methods provides the representation (7.22) for output of computed interval results in what is called the <u>A-format</u>.

In many applications, relative or percentage error is of more interest than absolute error. In this case, if 0 does not belong to the interval [a,b] (that is, ab
> 0), then one can take the <u>harmonic point</u> (or <u>mean</u>)

$$(7.23) \qquad x_H = 2ab/(a + b) = h([a,b])$$

as the approximation to x, and its relative error is bounded by

$$(7.24) \qquad |x - x_H|/|x| \le (b - a)/|a + b| = \frac{1}{2} \cdot w([a,b])/|m([a,b])|.$$

Multiplication of (7.24) by 100 provides the corresponding bound for the percentage
error. As before, a representation of the interval [a,b] can be obtained in terms of
the harmonic point x_H and the bound (7.24) for the relative error. One gets

$$(7.25) \qquad a = x_H/(1 + r([a,b])), \quad b = x_H/(1 - r([a,b]),$$

where

$$(7.26) \qquad r([a,b]) = (b - a)/(a + b) = \frac{1}{2} \cdot w([a,b])/m([a,b]).$$

The <u>R-format</u> for printing intervals [28], [39], consists of

$$(7.27) \qquad h([a,b]), \quad r([a,b]) .$$

(The sign of r([a,b]) is the same as the sign of h([a,b]), so the reconstruction of
[a,b] given by (7.25) is correct in either case.) For percentage error, the corresponding <u>P-format</u> is

$$(7.28) \qquad h([a,b]), \quad 100 \cdot r([a,b]) ,$$

where the interval [a,b] is checked to see that it does not contain zero before (7.26) or (7.27) is printed. On the other hand, the A-format output

(7.28)
$$m([a,b]), \quad \frac{1}{2} \cdot w([a,b]),$$

is defined for arbitrary intervals [a,b]. For additional information and methods of representing intervals and error bounds, see [75].

In the actual output of the programs [28], [39], an interval [a,b] is output in the form (a / b) in standard interval format, m A ε in A-format, h R r in R-format, and h P p in P-format, where

(7.29)
$$\varepsilon = \frac{1}{2} \cdot w([a,b]), \quad r = r([a,b]), \quad p = 100 \cdot r([a,b]),$$

are the measures of absolute, relative, and percentage error introduced previously. Thus, the value [4.442062 , 4.4420836] obtained as the interval value for a numerical integral in [28] is expressed variously as

(7.30)

$$(.44420620+01 \ / \ .44420836+01)$$
$$.444207280+01 \text{ A } .000001080+01$$
$$.44420728+01 \text{ R } .24780377-05$$
$$.44420728+01 \text{ P } .24780377-03 .$$

An extra digit has been added to the A-format so that if the last digit of b - a is odd, assuming that a,b have the same exponent, then the last digits of m([a,b]) and ε will each be equal to 5, so that the output interval (a / b) can be reconstructed exactly from (7.22) by a hand calculation. It is not clear exactly why the slash / was adopted as a "delimiter" to separate the lower and upper endpoints of intervals in the printing of the standard output [4]. A comma, as in ordinary interval nota- tion, or some other symbol without an operational significance is definitely prefer- able. Similarly, $\pm$ could replace A in the A-format, and % could replace P in the P- format. R or perhaps ® is a satisfactory indicator for relative error.

Calculations performed in interval arithmetic provide rigorous (if at times pes- simistic) limits within which the results of any exact computations have to lie. In many applications, such as the design of critical parts of an aircraft or spacecraft, guaranteed limits for the error of the computation may be required, even if expen- sive. The classical interpretation of interval analysis is as a means to provide such guaranteed error bounds for the unattainable results of exact calculations. An- other interpretation is that some problems have natural formulations in terms of in- tervals, and that the intervals obtained by computation are the answers to these prob- lems if properly posed, and not simply constructs which contain some mythical real quantity called the exact solution. In this view, interval analysis is a branch of mathematics (see, for example, [9]) which, like real and complex analysis, has its own theory, techniques, and domain of applicability to practical problems. For ex- ample, in determining the safety of an off-shore drilling platform, the strengths of the materials used are only known within limits (that is to say, within intervals).

Computationally, by applying forces which are again interval-valued to the structure, the resulting interval values of deflections can be taken to be the desired picture of the worst possible situation in storms of various strengths.

Whichever interpretation is taken of interval analysis, it provides a useful tool to the numerical analyst, as the programs discussed in the subsequent chapters will show. These follow the line of the classical interpretation, in which interval computations (combined with differentiation, where appropriate), provide rigorous bounds for quantities needed, and permit verification of the hypotheses of mathematical theorems by machine computation, rather than by tedious or even impracticable hand checking. Other insights into the use of the combination of automatic differentiation and interval analysis for automation of error estimation can be found in the paper by D. Stoutemeyer [97].

3. _Automatic computation of Lipschitz constants_. The first application of what can be called automatic mathematical analysis which will be examined here is the finding of _Lipschitz constants_ for functions (or operators). First of all, a function f of a single variable is said to be _Lipschitz continuous_ in an interval [a,b] if a constant L > 0 exists such that

$$(7.31) \qquad |f(x) - f(z)| \leq L \cdot |x - z|, \quad x,z \in [a,b].$$

A constant L satisfying (7.31) is naturally called a _Lipschitz constant_ for f on the interval [a,b]. If f is differentiable in [a,b], then, by the "Fundamental Theorem of Calculus",

$$(7.32) \qquad f(x) - f(z) = \int_{z}^{x} f'(t)dt .$$

Thus, for

$$(7.33) \qquad L = \sup_{a \leq t \leq b} \{|f'(t)|\} ,$$

one has that L is finite if f' is bounded on [a,b], and hence f is Lipschitz continuous on [a,b] in this case, since (7.31) then follows from (7.32). Now, if the complete derivative code list for f' has been formed and evaluated in interval arithmetic for the input interval X = [a,b], then the result is the evaluation of an interval extension F'(X) of f'(x). Of course, F'(X) = [c,d], where c,d are finite if the interval evaluation of F' can be carried out, and, from (7.14),

$$(7.34) \qquad L \leq |F'(X)| = \max\{|c| , |d|\} ,$$

so from (7.31),

$$(7.35) \qquad |f(x) - f(z)| \leq |F'(X)| \cdot |x - z| ,$$

so that the number F'(X) , which can be calculated automatically (see Note 7.1), is also a Lipschitz constant for f on [a,b]. (Actually, even the number L defined by (7.33) may not be the smallest possible constant for which f is Lipschitz continuous;

in any case, replacing one Lipschitz constant by another one which is larger does not disturb the sense of inequality (7.31).) It is also worth noting that the automatically obtained Lipschitz constant

(7.36) $$\Lambda = \left| F'(X) \right|$$

is rigorous, since the effect of roundoff error in the evaluation of $F'(X)$ is simply to decrease its lower endpoint c, and increase its upper endpoint d, and thus only make Λ larger than it would be if the arithmetic operations and library functions involved could be computed exactly. This means that it is logically correct to use the value of the Lipschitz constant (7.36) for f on $[a,b]$ obtained in this constructive manner to verify the hypotheses of mathematical theorems which require a value of this constant. The implication of this remark is that much painstaking checking of hypotheses of theorems in analysis can be automated. In particular, some of the tedious chores of numerical analysis in the verification of existence of solutions of equations and obtaining guaranteed error bounds can be performed by the same computer which is producing approximate solutions. As mentioned above, interval arithmetic has the capability of taking into account the effects of errors in the data and the roundoff errors in the actual computation by producing an interval which contains all real numbers which would be the result of exact operations on any exact value of the data lying in the given input data intervals. Quantities such as the automatic Lipschitz constant (7.36) make possible bounding of the truncation error, since from (7.35), one has

(7.37) $$\left| f(x) - f(z) \right| \le \left| F'(X) \right| \cdot w([a,b]) = \Lambda \cdot (b - a)$$

for $x, z \in X = [a,b]$.

Bounds for truncation error will be illustrated in the two subsequent chapters, but first it should be noted that the above statements apply immediately to vector intervals and operators, with absolute values (see (7.15)-(7.19)) replaced by norms. Thus, with the interpretation that $x = (x_1, x_2, \ldots, x_n)^T \in R^n$, and that $f: D \subset R^n \to R^m$, then $f(x) = (f_1(x), f_2(x), \ldots, f_m(x))^T \in R^m$, and $f'(x) = (\partial f_i(x)/\partial x_j)$ is an $m \times n$ matrix, and so on. The generalization of (7.35) obtained in this setting is

(7.38) $$\| f(x) - f(z) \| \le \| F'(X) \| \cdot \| x - z \| , \quad x, z \in X,$$

and, by defining

(7.39) $$w(X) = \max_{(i)} \{ w(X_i) \}$$

for a vector interval $X = (X_1, X_2, \ldots, X_n)^T$, it follows that

(7.40) $$\| f(x) - f(z) \| \le \| F'(X) \| \cdot w(X), \quad x, z \in X.$$

As before, the Lipschitz constant $\Lambda = \| F'(X) \|$ can be obtained automatically, using (7.16). The norm $\| F''(X) \|$ of the Hessian operator can be used in the same way to obtain a Lipschitz constant for the derivative f' of f rigorously (see (7.17)) [39], [51], [71].

4. Use of differentials in sensitivity and error analysis. In the general case of a function $f: D \subset R^n \to R^m$, the corresponding differential vectors

(7.41) $\qquad dx = (dx_1, dx_2, \ldots, dx_n)^T, \quad df = (df_1, df_2, \ldots, df_m)^T,$

are related by

(7.42) $\qquad\qquad\qquad df = f'(x) \cdot dx ,$

where $f'(x) = (\partial f_i / \partial x_j)$ is, of course, the Jacobian matrix of f, and $\cdot$ denotes matrix by vector multiplication. It follows from the definition of the derivative that df may be used as an approximation to the difference

(7.43) $\qquad\qquad\qquad \Delta f = f(x + dx) - f(x) ,$

with accuracy which increases as the differential vector dx approaches the zero vector $0 = (0,0,\ldots,0)$ in R^n. The linear relationship (7.42) between dx and df is often called a linearization of the (generally nonlinear) relationship (7.43) between Δf and dx. Automatic differentiation thus provides a computational method of linearization, which is probably the most frequently used approximation method in applied mathematics. Newton's method, discussed previously in Chapter 6, §3, is a typical application of linearization.

If the differential vector dx is interpreted to represent changes or errors in the data, then the corresponding differential df is an approximation to the resulting change in the output result. This type of error analysis, carried out in real arithmetic, may give a satisfactory error indication. Examples of linearized error analysis, applied to arithmetic operations and other algorithms, have been given by Stummel [98]. Interval arithmetic in (7.42) gives, of course, a rigorous inclusion.

Another use of the Jacobian matrix $f'(x)$ is in connection with what is called sensitivity analysis. The amount Δf_i that the ith component of the function vector will change in response to a change dx_j in the jth variable, all other quantities considered to be constant, is

(7.44) $\qquad\qquad\qquad \Delta f_i \approx (\partial f_i / \partial x_j) \cdot dx_j .$

Thus, the ijth element

(7.45) $\qquad\qquad\qquad f'(x)_{ij} = \partial f_i / \partial x_j ,$

$i = 1,2,\ldots,m; \; j = 1,2,\ldots,n$, can be taken as a measure of the response of $f_i(x)$ to a change in x_j, and is sometimes called the ijth sensitivity coefficient of the transformation $y = f(x)$. Automatic differentiation can provide these values, also.

A method for error analysis presented by Bauer [96] (see also [74]), makes use of the relative differential ρf of f, which is defined by

(7.46) $\qquad\qquad\qquad \rho f = df/f, \quad f \neq 0.$

Automatic differentiation could be used to obtain df, after which ρf can be found simply by division by f, which is of course assumed to be nonzero. It is also possible to obtain relative differentials directly. In fact, the analysis of Bauer [96] is based on the Kantorovich graph of the calculation, so that automation of the construction of the relative differential can be carried out on the basis of the corresponding code list. Using the notation,

(7.47)   RHOF = ρF = DF/F,

the relative differentials of the arithmetic operations are given in Table VII.2. By addition of similar formulas for library functions, relative differential code lists can be formed in the same way as the differential code lists discussed previously.

TABLE VII.2. RELATIVE DIFFERENTIALS

Line in Code List	Relative Differential of Label
(7.48) T = U + V	RHOT = (U*RHOU + V*RHOV)/T;
(7.49) T = U - V	RHOT = (U*RHOU - V*RHOV)/T ;
(7.50) T = U*V	RHOT = RHOU + RHOV;
(7.51) T = U/V	RHOT = RHOU - RHOV .

5. <u>Projects for Chapter 7</u>. In the following projects, assume that TYPE INTERVAL is available, if necessary. Code the required subroutines if it is not.

1°. Code the automatic calculation of Lipschitz constants for functions, matrices, and bilinear operators.

2°. Code linearized error analysis in real and interval arithmetic.

3°. Code the calculation of an arbitrary sensitivity coefficient of a given transformation.

4°. Extend Table VII.2 to the basic set of library functions used in CODEX, and automate the calculation of relative differentials by forming relative differential code lists, or subroutines for relative differentials.

CHAPTER VIII

SOLUTION OF NONLINEAR SYSTEMS OF EQUATIONS

Given an operator $f: D \subset R^n \to R^n$, the equation

(8.1) $$f(x) = 0$$

is a concise way in which to write the system of equations (6.15). In this chapter,
it will be assumed that f is a nonlinear operator, although a number of the results
given apply to the linear case as well.

1. <u>Simple iteration and the automatic contraction mapping theorem.</u> A problem
which is closely related to the solution of equations of the form (8.1) is the <u>fixed-
point problem</u> for an operator $\phi: D \subset R^n \to R^n$, which is to solve the equation

(8.2) $$x = \phi(x) ,$$

which again is concise notation for a system of n equations in n unknowns of the form

$$x_1 = \phi_1(x_1, x_2, \ldots, x_n),$$

(8.3) $$x_2 = \phi_2(x_1, x_2, \ldots, x_n),$$

$$\cdots \quad \cdots \quad \cdots \quad \cdots \quad \cdots$$

$$x_n = \phi_n(x_1, x_2, \ldots, x_n) .$$

The use of <u>iteration</u> to try to solve (8.2) for a fixed point of ϕ suggests itself
immediately. An <u>initial approximation</u> $x^0 = (x_1^0, x_2^0, \ldots, x_n^0)^T$ to x is chosen, and an
<u>improved approximation</u> x^1 (which will be "better" than x^0 only under certain condi-
tions) is calculated by

(8.4) $$x^1 = \phi(x^0) .$$

Continuation of this process with x^0 replaced by x^1 as the new initial approximation,
and so on, leads to the sequence $\{x^k\}$ of <u>iterates</u> of x^0 defined by

(8.5) $$x^{k+1} = \phi(x^k), \quad k = 0,1,2,\ldots .$$

This is the method of <u>simple iteration</u>, or <u>successive substitutions</u>. Under conditions
to be presented below, the sequence $\{x^k\}$ constructed in this way will converge to a
fixed point $x = x^*$ of ϕ.

Fixed point problems (8.2) can arise directly in numerical analysis, or by the
transformation of an equation of the form (8.1) into a fixed point problem. The in-
tent of such a transformation is to use the iteration method (8.5) for the solution
of equation (8.1); hence, in this case, ϕ is called an <u>iteration operator</u> for f.

For example, for ϕ defined by

(8.6)
$$\phi(x) = x - [f'(x)]^{-1}f(x) ,$$

the iteration method (8.5) is simply Newton's method (6.41)-(6.42), since

(8.7)
$$\delta x^k = - [f'(x^k)]^{-1}f(x^k) ,$$

and ϕ is called the <u>Newton operator</u> for f. (In actual computation, of course, it is more efficient to solve the linear system of equations (6.41) for δx^k than it is to invert the n×n matrix $f'(x^k)$ [23]. However, the representation (8.7) for δx^k is useful for theoretical purposes, as will be seen later.)

The classical contraction mapping theorem [71] gives conditions under which the existence of a fixed point x* of ϕ and the convergence of the sequence $\{x^k\}$ to x* are guaranteed; furthermore, error bounds are given for $\|x* - x^k\|$ in terms of a norm for R^n. In order to see how the application of this theorem can be automated by the methods of this book, a standard formulation of it will be analyzed. The set

(8.8)
$$\overline{U}(x^0,\rho) = \{x \mid \|x - x^0\| \le \rho\}$$

will be called, as usual, the <u>closed</u> <u>ball</u> <u>with</u> <u>center</u> x^0 and <u>radius</u> ρ in the space R^n.

The Classical Contraction Mapping Theorem [71]. If $\alpha < 1$ exists such that

(8.9)
$$\|f(x) - f(z)\| \le \alpha \cdot \|x - z\|$$

for all $x,z \in \overline{U}(x^0,\rho)$, and

(8.10)
$$\rho \ge \|x^1 - x^0\|/(1 - \alpha) = r,$$

then a fixed point x* of ϕ exists in the ball $\overline{U}(x^0,r)$.

Thus, if the hypotheses of this theorem are satisfied, then the <u>existence</u> of an $x = x*$ satisfying (8.2) is guaranteed; furthermore, the <u>error</u> <u>bound</u>

(8.11)
$$\|x* - x^0\| \le r$$

is provided for the initial approximation x^0 to x*. In the proof of this theorem, see [71], for example, it is also established that the sequence $\{x^k\}$ of iterates generated by (8.5) converges to x*, with

(8.12)
$$\|x* - x^k\| \le \alpha^k \cdot r, \quad k = 0,1,2,\ldots,$$

a result which has computational as well as theoretical implications.

It should be emphasized that the above theorem applies for general norms, not just for the maximum norm (6.30). There are, however, several criticisms which can be directed at this theorem from a practical standpoint:

1°. A Lipschitz constant α for ϕ is needed; this may not be easy to find.

2°. The value of ρ is not specified. If ρ is taken to be too large, then there may be no Lipschitz constant α of ϕ satisfying $\alpha < 1$; on the other hand, if ρ is taken too small, then the inequality (8.10) may not be satisfied. (In the case of systems

of polynomial equations, an explicit procedure for the choice of ρ has been worked
out by Rall [69].)

3°. In actual computation, x^1 (and likewise x^2, x^3, ...) are not obtained exactly, so that an error analysis is required to conclude that (8.10) is satisfied on
the basis of the computed value $\hat{x}^1 = x^1 + \epsilon^1$.

The first and third criticisms can be answered directly by the use of interval
arithmetic, combined with automatic differentiation. If the norm (6.30) (the maximum
norm) is used, then the set $\overline{U}(x^0,\rho)$ is also an interval vector. The midpoint $m(X)$
of the interval vector

(8.13) $X = (X_1,X_2,\ldots,X_n) = ([a_1,b_1],[a_2,b_2],\ldots,[a_n,b_n])$

is defined naturally to be the real vector

(8.14) $m(X) = (m(X_1),m(X_2),\ldots,m(X_n)) = \frac{1}{2}\cdot(a_1+b_1,a_2+b_2,\ldots,a_n+b_n)$,

and thus x^0 is the midpoint of (8.8), which can be represented as

(8.14) $\overline{U}(x^0,\rho) = x^0 + ([-\rho,\rho],[-\rho,\rho],\ldots,[-\rho,\rho]) = x^0 + \rho[-1,1]e$,

where [73],

(8.16) $e = (1,1,\ldots,1)$.

An interval of the form (8.15) is sometimes also called a cube. Since the ball
(or cube) (8.15) is an interval vector in the maximum norm, it will be denoted by
x^0_ρ, that is,

(8.17) $x^0_\rho = \overline{U}(x^0,\rho)$.

It follows that the Lipschitz constant

(8.18) $\alpha = \alpha(x^0_\rho) = \| \Phi'(x^0_\rho) \|$

can be obtained directly for given x^0 and ρ by the use of an interval extension Φ'
of ϕ'. Of course, ϕ' can be produced by automatic differentiation of ϕ, and evaluation of the result in interval arithmetic will give Φ' automatically.

The use of an interval extension Φ of ϕ will resolve the third difficulty, since
$\Phi(x^0)$ will be an interval vector. (There is no loss of generality in assuming that
x^0 is an exact machine vector, and the endpoints of $\Phi(x^0)$ are also.) Thus, $\Phi(x^0)$ contains the result $\phi(x^0)$ of an exact transformation of the vector x^0 by the operator ϕ.
It follows that

(8.19) $\| \Phi(x^0) - x^0 \| \geq \| \phi(x^0) - x^0 \| = \| x^1 - x^0 \|$,

which makes possible the rigorous verification of (8.10). These observations can be
combined to give the following result.

The Automatic Contraction Mapping Theorem. If

(8.20)
$$\alpha = \| \Phi'(X_\rho^0) \| < 1$$

and

(8.21)
$$\rho \geq \| \Phi(x^0) - x^0 \|/(1 - \alpha) = r,$$

for interval extensions Φ' of ϕ' and Φ of ϕ, respectively, then a fixed point x* of ϕ exists in X_r^0.

The hypotheses of this theorem are <u>computationally verifiable</u>, given x^0 and ρ, or, what is the same, X_ρ^0. Furthermore, all that is required is essentially the computation of the Jacobian ϕ' of ϕ, and its evaluation in interval arithmetic to obtain Φ'. The error bound (8.11) also follows, of course, if (8.20) and (8.21) are satisfied. Thus, this theorem can be used to guarantee existence of a fixed point of ϕ and provide an error bound for the initial approximation on the basis of an actual computation, so that it is reasonable to call it an <u>automatic</u> theorem, provided that automatic differentiation is used to obtain the necessary derivatives to evaluate the interval Jacobian matrix needed. Furthermore, the satisfaction of (8.20) and (8.21) guarantees the convergence of the iteration process (8.5), which can be useful in the investigation of various iteration operators ϕ, as well as in a specific computation. It is not difficult to construct software to implement the automatic contraction mapping theorem, and the results can be used for a number of practical and theoretical purposes.

The problem of the choice of X_ρ^0 remains in the cases not covered by special results, such as the ones known for polynomial systems of equations [69]. There are several strategies which may be adopted for this purpose, such as:

1°. Choose initially a cube X which is fairly certain to contain a solution x* of (8.2), based perhaps on some outside information about the problem, and then take

(8.22)
$$x^0 = m(X), \quad \rho = \frac{1}{2} \cdot w(X) ,$$

so that $X_\rho^0 = X$, and hope for the best.

2°. As is more usual, choose a point x^0 which is believed to be an approximation to x* with some accuracy, then compute $\Phi(x^0)$ and

(8.23)
$$\eta = \| \Phi(x^0) - x^0 \| ,$$

and then try $\rho = k \cdot \eta$ for one or more values of $k > 1$ to see if (8.20) and (8.21) are satisfied, decreasing k if α is too large, or increasing k if ρ is too small. In the next section, it will be shown that $k = 2$ is satisfactory if ϕ is the Newton iteration operator (8.6).

2. <u>Newton's method and the automatic Kantorovich theorem</u>. The use of the iteration operator (8.6) for the solution of equation (8.1) gives rise to the <u>Newton sequence</u> $\{x^k\}$ defined by

(8.24)
$$x^{k+1} = x^k - [f'(x^k)]^{-1} f(x^k), \quad k = 0,1,2,\ldots,$$

according to (8.5). The existence of a solution x* of equation (8.1) and the convergence of the sequence $\{x^k\}$ generated by (8.24) to x* can be established on the basis
of the famous theorem of L. V. Kantorovich [35] (see also [71]), given here in the
very neat formulation of Ortega [61].

The Classical Kantorovich Theorem (Ortega [61]). Given $\overline{U}(x^0,\rho)$, suppose that
$[f'(x^0)]^{-1}$ exists, and

(8.25) $$\| [f'(x^0)]^{-1} \| \le B_0, \quad \|x^1 - x^0\| \le \eta_0 ,$$

and f' is Lipschitz continuous in $\overline{U}(x^0,\rho)$ with Lipschitz constant K, that is,

(8.26) $$\|f'(x) - f'(z)\| \le K \cdot \|x - z\|, \quad x,z \in \overline{U}(x^0,\rho) ;$$

then, if

(8.27) $$h_0 = B_0 \eta_0 K \le \frac{1}{2} \quad \text{and} \quad \rho \ge (1 - \sqrt{1 - 2h_0}) \eta_0 / h_0 = r ,$$

a solution x* of equation (8.1) exists in the ball $\overline{U}(x^0,r)$.

As in the case of the classical contraction mapping theorem, several remarks are
in order about this theorem as examined from a practical standpoint.

1°. The condition on the invertibility of $f'(x^0)$ is a natural one, since it is
necessary to calculate x^1 to get the Newton sequence started. However, since this
cannot be done exactly in general in actual computation, an error analysis is required
to establish that one can find B_0, η_0, for which (8.25) can be guaranteed to hold rigorously.

2°. The Lipschitz constant K appearing in (8.26) must also be obtained in some
rigorous fashion. In the original formulation of the above theorem by Kantorovich
[35] (see also [71]), K was taken to satisfy

(8.28) $$K \ge \sup_{x \in \overline{U}(x^0,\rho)} \{\|f''(x)\|\} ,$$

and thus is a Lipschitz constant for f', as in Chapter 7.

3°. The choice of ρ, given x^0, is not difficult in the case of this theorem.
As pointed out by Kantorovich [35], it is sufficient to calculate x^1 and then take

(8.29) $$\rho = 2\eta_0 ,$$

and any rigorous upper bound for this quantity will do.

The comments in remark 1° can be dealt with as in the case of the contraction
mapping theorem, if interval extensions F, F' of f, f', respectively, are used, once
again combining automatic differentiation to produce code to be executed in interval
arithmetic. The interval matrix $F'(x^0)$ can be inverted by Hansen's method [31], and,
since

(8.30) $$x^1 - x^0 \in - [F'(x^0)]^{-1} F(x^0) ,$$

one can use the rigorous values

(8.31) $$B_0 = \|[F'(x^0)]^{-1}\|, \quad \eta_0 = \|[F'(x^0)]^{-1}F(x^0)\|,$$

obtained by interval computation as the bound required in (8.25). A value for the Lipschitz constant K can be computed in a similar fashion. As before, the ball $\overline{U}_0 = \overline{U}(x^0, 2\eta_0)'$ is identified with the cube $X^0_{2\eta_0}$, in which a Lipschitz constant K which satisfies (8.28), given an interval extension F" of f", is

(8.32) $$K = \|F''(X^0_{2\eta_0})\|,$$

once again rigorously guaranteed. With the above values, the following theorem holds.

The Automatic Kantorovich Theorem. If for the values B_0, η_0, given by (8.31) and K given by (8.32), one has

(8.33) $$h_0 = B_0\eta_0K \le \frac{1}{2},$$

then a solution x* of equation (8.1) exists in the cube X^0_r, where

(8.34) $$r = ((1 - \sqrt{1 - 2h_0})/h_0)\eta_0 \le 2\eta_0.$$

In addition to guaranteeing the existence of a solution x* of (8.1) under satisfaction of its hypotheses, the above theorem provides the error bound

(8.35) $$\|x^* - x^0\| \le r = ((1 - \sqrt{1 - 2h_0})/h_0)\eta_0$$

for x^0 as an approximation to x*. From a theoretical standpoint, this theorem also establishes the convergence of the Newton sequence $\{x^k\}$ to x*. The convergence is quadratic if h < 1/2 [35], [61], [71]; in fact, for

(8.36) $$\theta = (1 - \sqrt{1 - 2h_0})/(1 + \sqrt{1 - 2h_0}),$$

Gragg and Tapia [24] have shown that

(8.37) $$\|x^* - x^k\| \le \theta^{2^k}\eta_0/(\sqrt{1 - 2h_0}(1 - \theta^{2^k})), \quad k = 0,1,2,\ldots.$$

This automatic (or interval) Kantorovich theorem has a place in history as the first existence theorem for solutions of systems of nonlinear equations to be completely automated by using software for automatic differentiation and interval arithmetic, first in 1967 [26], [27], [71], followed later in 1972 by a completely rigorous version [39]. One drawback of the automatic Kantorovich theorem is that two relatively expensive operations are required in this calculation, the inversion of the interval Jacobian matrix $F'(x^0)$ and the computation of the interval Hessian operator to obtain the bounds (8.31) and (8.32), respectively. It has been shown by Rall [73] that a recent theorem of Moore [54], to be described in the next section, can be used to establish existence of solutions of nonlinear equations automatically without interval matrix inversion or construction of the Hessian operator being required, under essentially the same conditions in which the Kantorovich theorem is applicable. The second disadvantage of the automatic Kantorovich theorem is that its hypotheses have to be verified in a cube, rather than a more general interval, which is also a limitation overcome in the methods to be discussed in the next section.

3. <u>Interval versions of Newton's method and the automatic theorems of Nickel and Moore</u>. This section is devoted to a description of interval iteration methods, for which the initial data required is an interval x^0 which is thought to contain a solution x^* of equation (8.1), rather than an approximate value x^0 of x^*. This initial interval is not restricted to be a cube, which is advantageous if the various variables have ranges of values which differ by orders of magnitudes, and perhaps also have entirely different interpretations in an applied problem. For example, in the optimization of the operation of a refinery, x_1 could represent barrels of oil, x_2 pressure in pounds per square inch, x_3 temperature in degrees Fahrenheit, x_4 a flow rate in gallons per minute, x_5 cost of feedstock and energy input in dollars, x_6 time in hours, and so on. Thus, it would be difficult to interpret the vector $x = (x_1, x_2, x_3, x_4, x_5, x_6, \ldots)$ as some kind of homogeneous quantity as in the case all components have similar interpretations as distances, costs, or other related values. Also, a requirement that all variables have the same absolute ranges of variation as the largest variable, which is the case in a cube, could furthermore mean that the cube could extend beyond the domain of definition D of the function or derivative being computed, thus causing computational difficulties. Therefore, intervals may be more natural regions to work with in some problems than provided by balls in normed linear vector spaces, as considered in the standard approach to the solution of operator equations via functional analysis [62], [71]. Since <u>analytic</u> existence theorems, such as the classical contraction mapping and Kantorovich theorems, are formulated in terms of balls in normed linear spaces, some scaling of variables (or alteration of the norm) may be required before these theorems can be applied to the problem at hand. These transformations needed to scale all the variables into numbers of about the same magnitude may be avoided in many cases by the use of interval existence theorems, with all variables allowed to take on values in their "natural" ranges.

Several interval existence theorems will be described in this section, all of which have been automated and implemented in the program [39], in addition to the automatic Kantorovich theorem discussed in the previous section. The first automatic theorem to be presented here, however, is one that can be used to establish <u>nonexistence</u> of solutions x^* of (8.1) in the initial interval x^0.

The Automatic Nonexistence Theorem. If X is a given interval, and F is an interval extension of f, then there is no solution $x = x^*$ of (8.1) in X if

(8.38) $0 \bar{\in} F(X)$.

This is simply the contrapositive of the assertion that if $x^* \in X$ and $f(x^*) = 0$, then $0 \in F(X)$ for an arbitrary interval extension F of f, by definition. Thus, the interval $F(x^0)$ should be checked to see that it contains 0 before any further effort is made to find x^* in x^0. Of course, if F(X) extends beyond f(X), which is to be expected, then it does <u>not</u> follow from $0 \in F(x^0)$ that $f(x*) = 0$ for some $x* \in x^0$.

The first interval version of the Newton iteration to be considered is the <u>in-terval Newton method</u>, formulated by Moore [53] and investigated further by Nickel [59]. This method starts from an initial interval x^0, and generates the sequence of intervals $\{x^k\}$ by first computing

$$(8.39) \qquad z^k = m(x^k) - [F'(x^k)]^{-1}F(m(x^k)),$$

and then taking

$$(8.40) \qquad x^{k+1} = x^k \cap z^k, \quad k = 0,1,2,\ldots .$$

In actual practice, this process will stop with $x^{k+1} = \emptyset$, the empty set, or with $x^{k+1} = x^k$, since there are only a finite number of endpoints available in the number set of a computer. The following theorem gives an existence (or nonexistence) condi-tion which can be verified automatically.

An Automatic Existence or Nonexistence Theorem for the Interval Newton Method (Nickel). If

$$(8.41) \qquad z^0 \subset x^0 ,$$

then x^0 contains a solution $x*$ of equation (8.1); on the other hand, if

$$(8.42) \qquad x^0 \cap z^0 = \emptyset ,$$

the empty set, then x^0 does not contain a solution $x = x*$ of the equation $f(x) = 0$.

A proof of this theorem can be found in [59]. The theorem is completely auto-matic in character, since

$$(8.43) \qquad z^0 = m(x^0) - [F'(x^0)]^{-1}F(m(x^0))$$

is computable immediately for a given x^0 by the use of software for differentiation and interval computations. However, (8.43) does require the inversion of an interval matrix, which is a fairly extensive calculation. The iteration process (8.39)-(8.40) is available in the program [39] as an option. Compared to the Kantorovich theorem, the above theorem does not require the computation of the Hessian operator, but no comparison of the ranges of validity of the two theorems along the lines of [73] is known to have been derived to date.

A second interval iterative method which is available as an option in the pro-gram [39] is the <u>Krawczyk method</u> [38]: Here, given the initial interval x^0, one com-putes

$$(8.44) \qquad z^0 = m(x^0) - [f'(m(x^0))]^{-1}f(m(x^0)) +$$
$$+ (I - [f'(m(x^0))]^{-1}F'(x^0))(x^0 - m(x^0)) ,$$

and then

(8.45)
$$x^1 = x^0 \cap z^0 ,$$

a process which can be iterated to produce a sequence of intervals x^k, $k = 0,1,2,...,$ as before. (In (8.44), I denotes the $n \times n$ identity matrix.) An examination of (8.44) reveals that it is highly economical with respect to interval calculations; only the interval Jacobian $F'(x^0)$ is needed, and the multiplication of an interval matrix with the interval $x^0 - m(x^0)$ is the only other interval calculation of any extent. The matrix $f'(m(x^0))$ which is inverted in the formula has real coefficients, so no interval matrix inversions are required. Since the software for the computation of (8.44) was made available in 1972 by Kuba and Rall [39], it was ready for use in 1977 when Moore [54] published the following result.

Moore's Existence Theorem. If X is a given interval, $y \in X$, the operators f and its derivative f' have interval extensions F and F', respectively, and the interval K(X) defined by the Krawczyk transformation

(8.46)
$$K(X) = y - Yf(y) + (I - YF'(X))(X - y) ,$$

where Y is a nonsingular real matrix, has the property

(8.47)
$$K(X) \subset X ,$$

then a point $x^* \in X$ exists such that $f(x^*) = 0$.

One sees immediately that z^0 as defined by (8.44) is the Krawczyk transformation of x^0 obtained by taking

(8.48)
$$y = m(x^0), \quad Y = [f'(m(x^0))]^{-1} .$$

This gives the following theorem, which was added as an option to the program [39] by Mrs. Julia Gray.

The Automatic Moore Existence Theorem No. 1. If for z^0 computed by (8.44), one has

(8.49)
$$z^0 \subset x^0 ,$$

then the given interval x^0 contains a solution $x = x^*$ of the equation $f(x) = 0$.

It is assumed that the real matrix inversion is done carefully enough and monitored to assure that the matrix obtained as the inverse is indeed the inverse of some real matrix (hopefully close to the matrix being inverted).

By making various choices of the point y and the matrix Y in the Krawczyk transformation (8.46), other automatic versions of the theorem of Moore can be obtained. In the above, the Jacobian of f at the midpoint of x^0, namely, $f'(m(x^0))$, was inverted to obtain Y. Another choice would be the midpoint of the (interval) Jacobian $F'(X^0)$, that is, the real matrix $m(F'(X^0))$, to invert to obtain Y, as used in the paper of Moore and Jones [56] on safe starting intervals for iteration methods. Here, the midpoint of an interval matrix

(8.50)
$$M = (M_{ij}) = ([a_{ij}, b_{ij}])$$

is, of course, the real matrix

(8.51) $$m(M) = (m([a_{ij}, b_{ij}]) = \frac{1}{2}(a_{ij} + b_{ij}) ,$$

with coefficients which are the midpoints of the interval coefficients of the interval matrix. Since $F'(X^0)$ has to be computed as part of the Krawczyk transformation of X^0, it is economical to take its midpoint and invert to obtain Y, provided that $m(F'(X^0))$ is "safely" invertible, and thus use the choices

(8.52) $$y = m(X^0), \quad Y = [m(F'(X^0)]^{-1} .$$

This option has also been added to the program NEWTON, and furnishes a second automatic theorem.

The Automatic Moore Existence Theorem No. 2. If Z^0 is computed by

(8.53) $$Z^0 = m(X^0) - [m(F'(X^0)]^{-1}f(m(X^0)) +$$

$$+ \quad (I - [m(F'(X^0)]^{-1}F'(X^0))(X^0 - m(X^0)) ,$$

and $Z^0 \subset X^0$, then the interval X^0 contains a solution $x*$ of the equation (8.1).

There is also a nonexistence result based on the transformation (8.46).

The Automatic Moore Nonexistence Theorem [54]. If

(8.54) $$K(X) \cap X = \emptyset,$$

where $K(X)$ is defined by (8.46), then the interval X does not contain a solution $x*$ of the equation $f(x) = 0$.

Rall [73] has shown that the two automatically verifiable theorems of Moore have essentially the same region of applicability as the theorem of Kantorovich in the sense that if x^0 is actually a good approximation to a solution $x*$ of (8.1), then the conclusions of all three theorems will be positive. This has the highly practical implication that interval matrix inversion and interval evaluation of the Hessian operator (or even coding the second derivatives) are not required, and the elaborate program [39] of Kuba and Rall can be simplified drastically to obtain an efficient execution of Newton's method with optional verification of existence and rigorous error estimation.

The interval methods presented above provide immediate error estimates once the solution $x*$ has been certified to belong to an interval X. For

(8.55) $$y = m(X) ,$$

one has, of course,

(8.56) $$\|x* - y\| \leq \frac{1}{2} \cdot w(X) .$$

As mentioned earlier, componentwise error bounds may be more significant than the overall bound (8.56). For $X = (X_1, X_2, \ldots, X_n)$, one has

(8.57) $$|x_i - y_i| \leq \frac{1}{2} \cdot w(X_i), \quad i = 1, 2, \ldots, n,$$

for the absolute errors, and by use of the harmonic points $h(X_i)$ instead of the mid-points of the intervals X_i, bounds for the relative or percentage errors can be obtained as in (7.23)-(7.27).

4. The program NEWTON. The program [39] provides to the user a number of options in the solution of systems of nonlinear equations using real (double precision, in this case) or interval arithmetic. Since automatic differentiation is used, the amount of input required is minimized. The user has to supply formulas, which can include shorthand variables, for the functions defining the system of equations (8.1), that is, for $f(x) = (f_1(x), f_2(x), \ldots, f_n(x))$, the n functions

(8.58)
$$f_i(x) = f_i(x_1, x_2, \ldots, x_n), \quad i = 1, 2, \ldots, n,$$

are supplied in the form Fi = Fi(X1,X2,...,Xn). An initial approximation x^0 to x^*, or an initial interval x^0 thought to contain x^* is also input, together with parameters indicating the mode of computation, the type of existence theorem to be applied, if desired, and various other controls on the computation and form and amount of output.

One basic mode of operation of the program is to simply generate the Newton sequence (8.24) in double precision arithmetic, starting from a given vector x^0, until preselected convergence or divergence criteria are met, or until a prescribed number of iterations has been performed. In the case that the Newton sequence has converged numerically according to one of the criteria given below, then the iterate x^K of x^0 obtained as the output of this calculation can be used as the initial approximation in the application of existence theorems and the computation of error bounds. In other words, if the numerical Newton sequence has apparently converged, the resulting approximate solution obtained has a good chance of being close to an exact solution x^*; at least, the attempt to verify the conditions of an existence theorem would appear to be warranted at this point.

Other modes of operation which have been implemented in the program are the interval Newton's method (8.39)-(8.40), and the Krawczyk method defined by (8.44) and (8.40), and the Moore-Jones method using (8.53) and (8.40). Since these have been described in the previous section (see also [38], [54], [56]), attention will be devoted here to the real version.

The numerical calculation of the Newton sequence (8.24) is carried out in the double precision real mode, if selected, until criteria for convergence or divergence given by the user are satisfied, until a given number of iterations are performed, or until some kind of a fault condition is encountered, for example, failure of $f'(x^k)$ to be invertible, or x^k lying outside of the domain of definition D of f or f', so that $f(x^k)$ or the Jacobian matrix $f'(x^k)$ cannot be evaluated. The magnitudes of the vector $f(x^k)$ and the kth increment (or correction)

(8.59)
$$\delta x^k = x^{k+1} - x^k, \quad k = 0,1,2,\ldots,$$

are monitored, and if

$$\text{(i)} \quad \text{CNVERG(1)} \geq \| f(x^k) \|, \text{ or}$$

(8.60)

$$\text{(ii)} \quad \text{CNVERG(2)} \geq \| \delta x^k \| \quad ,$$

where CNVERG(i), i = 1,2, are supplied by the user, then the calculated Newton sequence will be said to have <u>converged numerically</u>. (There is a third convergence parameter CNVERG(3), with default value of 0.5, which is used in connection with monitoring the value of h_0 in the Automatic Kantorovich Theorem. Since this theorem is now essentially only of historical interest, CNVERG(3) can be ignored. The default values of the other two convergence parameters in (8.60) are both 0.0, and are thus only active if set, in the large majority of cases.)

On the other hand, if for given divergence parameters DIVERG(i), i = 1,2,3, one observes,

$$\| f(x^k) \| > \text{DIVERG(1), or}$$

(8.61)

$$\| \delta x^k \| > \text{DIVERG(2), or}$$

$$\| [f'(x^k)]^{-1} \| > \text{DIVERG(3)} \quad ,$$

then the numerical Newton sequence is said to have diverged, and the iteration is terminated with a statement of the appropriate reason. If the numerical Newton sequence just flounders about, not exhibiting either convergence or divergence according to the criteria (8.60) or (8.61), respectively, then the iteration will be terminated at the number NSTEP specified by the user, with an indication that this is the reason for termination, rather than convergence or divergence.

In the favorable case that convergence is indicated by the satisfaction of the condition (8.60)(i), which is checked after the calculation of $f(x^k)$ and before δx^k is computed, then the output for the numerical Newton method will be taken to be x^k; similarly, if (8.60)(ii) holds, then x^{k+1} is available from $x^{k+1} = x^k + \delta x^k$, and is the result output by the program. It should be emphasized that the automatic existence theorems incorporated in the program [39] need only an initial approximation x^0 to the desired solution x* of (8.1), or an interval X^0 thought to contain x*. It is <u>not</u> necessary to obtain these approximations from the numerical Newton sequence. However, as the existence theorems are related in one way or another to Newton's method, since all require evaluation of the Jacobian matrix f'(x) and an interval extension F'(X) of it over some interval X, it does not seem to be a bad idea to investigate the behavior of the numerical Newton sequence starting at the given approximate solution (or midpoint of the initial interval) before going on to the application of an existence theorem, especially since an improved initial approximation will result if the Newton iteration converges numerically. The next section will be devoted to some additional comments about initial approximations.

A simplification of the program [39] for real calculation which suggests itself immediately is to avoid inversion of the matrix $f'(x^k)$, which can be done by solving the linear system (6.42) for δx^k directly, a far more efficient process. The inverse of the Jacobian could then be done only when needed for interval computation, or for theoretical purposes. Furthermore, if only approximate solutions are needed for the verification of existence, then the sequence calculated for this purpose could be obtained just as well using only single precision arithmetic. However, it is useful to have the double precision calculation available as an option in case refinement of approximate solutions is desired, particularly after the existence of a solution and the convergence of the Newton sequence to it have been verified.

5. Some methods for finding initial approximations. It can happen that it is not possible in certain problems to find initial approximations x^0 to a solution $x*$ of (8.1), or a region X^0 containing $x*$ in a convenient fashion. In these situations, it also turns out that there are some methods available which can be automated using software for automatic differentiation and interval calculations. For example, the bisection method of Moore and Jones [56] starts with a large interval X, and then determines in a finite number of steps that X either contains no solution $x*$, or finds a subinterval X^0 of X which contains a solution, and furthermore guarantees that some iteration method will generate a sequence starting from a point $x^0 \in X^0$ (such as the midpoint $m(X^0)$) which converges to $x*$. This method uses calculations of the form (8.53) to test subintervals for inclusion of solutions.

Another method which is capable of finding initial approximations suitable for computational and theoretical purposes is the so-called continuation (or homotopy) method. Here, an operator $H(x,t)$ is constructed in $R^n \times [0,1]$ such that

$$(8.62) \qquad\qquad H(x^0,0) = f(x^0),$$

and

$$(8.63) \qquad\qquad H(x,1) = f(x) .$$

The equation

$$(8.64) \qquad\qquad H(x,t) = 0$$

thus defines a homotopy curve $x(t)$ as a function of the artificial variable (or homotopy parameter) t, which is assumed to be a smooth arc connecting the known (but unsuitable) initial approximation $x^0 = x(0)$, and the unknown solution $x* = x(1)$ of the equation (8.1). Thus, if the homotopy curve can be followed closely enough, one can find an initial approximation of $x*$ of required accuracy.

A simple and effective procedure of this type has been described by Li and Yorke [40]. It is a more general form of the method of Davidenko [17], [70]. In either case, automatic differentiation can be used to find a system of ordinary differential equations for the homotopy curve $x = x(t)$. To illustrate this, in Davidenko's method, the variable t is introduced by setting

$$(8.65) \qquad\qquad H(x,t) = f(x) + (1 - t)f(x^0), \quad 0 \le t \le 1,$$

and differentiating to obtain the initial-value problem

$$(8.66) \qquad\qquad \frac{dx}{dt} = - [f'(x(t))]^{-1}f(x^0), \quad x(0) = x^0,$$

which is integrated numerically to obtain a good approximation to $x(1) = x^*$. The Jacobian matrices $f'(x(t))$ required in this process can be obtained by automatic differentiation.

6. Projects for Chapter 8.

1°. Write a compact program for Newton's method which solves linear equations to obtain δx^k in the real mode (single or double precision), and implements also the interval iteration $X^{k+1} = Z^k \cap X^k = K(X^k) \cap X^k$ for the Krawczyk operators (8.44) and (8.53) and the corresponding automatic existence and nonexistence theorems.

2°. Write a program to maximize or minimize the functional $f = f(x)$, using Newton's method to solve the gradient equations $\nabla f(x) = 0$. Here, it will be necessary to generate second derivatives automatically, since the Jacobian matrix of the gradient system is the Hessian operator $f''(x) = (\partial^2 f / \partial x_i \partial x_j)$.

3°. Write a program to implement Davidenko's method (8.66). Assume that software for solving the initial value problem for a system of ordinary differential equations is available.

4°. Write a program to implement the homotopy method of Li and Yorke [40], under the same assumption as in the previous project.

5°. Write a program for step-by-step (or discrete) continuation. Starting from $t_0 = 0$, one chooses $t_{k+1} = t_k + \delta t_k$, and then solves $H(x, t_{k+1}) = 0$ by Newton's method with $x = x^k$ as the initial approximation to get $x = x^{k+1}$, stopping when $t_k = 1$. How can δt_k be chosen automatically?

CHAPTER IX

NUMERICAL INTEGRATION WITH RIGOROUS ERROR ESTIMATION

This chapter is devoted to a description of some applications of automatically generated Taylor coefficients and series, in particular, the program INTE [28], which uses automatically generated Taylor coefficients of the integrand and interval arithmetic to perform a complete and rigorous error analysis of a numerical integration in an automatic fashion. Definitions of Taylor coefficients of a real function f and their relationships to derivatives of f have been given in Chapter 4 (see particularly equations (4.1)-(4.9)), and methods for their automatic generation have also been presented, based on the idea of calling subroutines for recursion formulas in the sequence specified by the function code list for f. There are numerous applications of software for this purpose; attention will be confined here to ones which have actually been implemented.

1. <u>Notation</u>. As in (4.1)-(4.2), the exact Taylor coefficients of a real function f at a point x will be denoted by

$$(9.1) \qquad f_i(x) = \frac{1}{i!}\frac{d^i f(x)}{dx^i} = \frac{1}{i!}f^{(i)}(x) ,$$

$i = 0,1,2,\ldots$, with the standard convention being adopted for $i = 0$, namely, $0! = 1$ and $f^{(0)}(x) = d^0 f(x)/dx^0 = f(x)$. It will be convenient to denote the Taylor coefficients obtained from a function code list for F at some point x by F(I), $I = 0,1,2,\ldots,K$, that is,

$$(9.2) \qquad F(I) = \frac{1}{I!}*F^{(I)}(x) , \quad I = 0,1,\ldots,K,$$

and the interval extensions of the coefficients (7.2) by the same symbols, or by

$$(9.3) \qquad F(I)(X) = F(I)([a,b]), \quad I = 0,1,\ldots,K,$$

to specify the evaluation of the interval extension of F(I) over the interval X = [a,b]. Ordinarily, it will be clear from the context whether interval or real arithmetic is being used in the computation.

Denoting the Taylor coefficients of F at some fixed point $x = x_0$ by $F_0(I)$, one can write the <u>Taylor polynomial</u> (4.5) as

$$(9.4) \qquad PKF(x) = \sum_{I=0}^{K} F_0(I)*(x - x_0)**I ,$$

which can be evaluated in either real or interval arithmetic, once the corresponding values of the Taylor coefficients F(I) have been computed for I = 0,1,...,K. The interval (4.8), that is, $[x_0,x]$ if $x \geq x_0$ or $[x,x_0]$ if $x \leq x_0$, will be denoted by X. Using interval arithmetic, the <u>interval</u> <u>remainder</u> <u>term</u> in the expansion of F(x) into Taylor series is defined to be

(9.5) $$RF(X) = F_0(K+1)(X) * (x - x_0) ** (K+1) .$$

It follows from (9.4) and (4.4)-(4.5) and either (4.6) or (4.7) that

(9.6) $$F(x) \in PKF(x) + RF(X) .$$

This provides a way to establish rigorous error bounds for the use of the Taylor polynomial $p_k f$ as an approximation to a function f which is differentiable at least k + 1 times. Since polynomials are easy to manipulate mathematically in the sense that they can be evaluated, integrated, differentiated, and so on, with relatively little effort, much of classical numerical analysis can be automated, including error analysis, by using the formulation (9.6). In particular, (9.6) provides a way to produce interval extensions of library functions needed for interval arithmetic.

It is no restriction to the method of Taylor series expansion if the function f = f(x,y,z,...) is expressed as a function of several variables, in which y, z, ... are also functions of x. The method of automatic generation of Taylor coefficients then gives the coefficients of f in terms of the Taylor coefficients of y, z, ..., since the derivatives of f can be expressed directly in terms of the derivatives of its variables by an extension of the formula (3.3).

2. <u>Numerical integration of systems of ordinary differential equations</u>. One of the most important applications of the automatic generation of Taylor coefficients, for which software was developed very early, is the numerical solution of the initial value problem for systems of ordinary differential equations. For details and examples, the reports, papers, and books of R. E. Moore should be consulted [5], [51], [52], [53], [55], [57], and also the descriptions of the program TAYLOR by A. Reiter [78], [80]. Only the basic idea will be outlined here, for a single function of one real variable. The same procedure extends immediately to vector-valued functions of one real variable, that is, to systems of ordinary differential equations with given initial conditions. This is often called the <u>initial-value</u>, or <u>Cauchy problem</u> for ordinary differential equations. Suppose that the given problem consists of the <u>differential equation</u> (DE):

(9.7) $$y'(x) = f(x,y(x)) ,$$

and the <u>initial condition</u> (IC):

(9.8) $$y(x_0) = y_0 .$$

Assuming that y(x) and f(x,y(x)) = f(x) have Taylor series expansions at $x = x_0$,

(9.9) $$y(x) = \sum_{i=0}^{\infty} y_i (x - x_0)^i, \quad f(x) = \sum_{i=0}^{\infty} f_i (x - x_0)^i ,$$

respectively, it follows from the differential equation (9.7) that the Taylor coefficients of y and f are related by

(9.10)
$$y_{i+1} = \frac{1}{i+1} \cdot f_i, \quad i = 0,1,2,\ldots .$$

From (9.8), y_0 is given, and since

(9.11)
$$f_0 = f(x_0, y(x_0)) = f(x_0, y_0) ,$$

then y_1 can be obtained from (7.10). Knowing (y_0, y_1), the execution of the subroutine call list for Taylor coefficients formed from the function code list for f will give the Taylor coefficients (f_0, f_1). The value of the Taylor coefficient f_1 computed in this way yields the Taylor coefficient y_2 of y from (9.10), and the resulting vector of coefficients (y_0, y_1, y_2) can be used to obtain (f_0, f_1, f_2), and so on, until as many coefficients as desired of the series (9.9) are computed. It is not at all necessary to assume that y is analytic, meaning that the infinite series in (9.9) converge in some disc centered on x_0 with positive radius. One can work under the assumption of differentiability $k + 1$ times of y, and use the Taylor polynomial with remainder term,

(9.12)
$$y(x) = \sum_{i=0}^{k} y_i (x - x_0)^i + R_k y(x; x_0) .$$

By use of the expression (9.5) for the remainder term and (9.10), one has for $x \geq x_0$,

(9.13)
$$R_k y(x; x_0) \in Y(k+1) \cdot (w(X))^{k+1} = \frac{F(k)}{k+1} \cdot (w(X))^{k+1} ,$$

so that recursive calculation of interval extensions of the Taylor coefficients of f over the interval X also yields bounds for the remainder term in the expansion (9.12) of y, given an interval bound Y for y on X. Moore observed that if the interval Y_0 contains the initial value y_0 of y properly, then $y(x)$ will be contained in Y_0 for $w(X) = x - x_0 = h$ sufficiently small, so that the performance of this calculation in interval arithmetic provides also automatic and rigorous error bounds for the round-off, truncation error, as well as error due to impreciseness of the initial data y_0 and coefficients of the function f [51], [53], [55]. In practice, the region of integration of the equation, say $[x_0, x_n]$, is broken into subintervals $X_i = [x_{i-1}, x_i]$, $i = 1, 2, \ldots, n$, of length $w(X_i) = w(X)/n = h$, and thus the Taylor series method has an order of convergence over the entire interval which is equal to k, that is, the error between an exact solution y and the midpoints of the interval solutions actually computed will be of order $h^k = (w(X)/n)^k$.

3. <u>Numerical integration</u>. The problem of <u>mechanical</u> (or <u>numerical</u>) <u>quadrature</u>, or simply <u>numerical integration</u> is to find the value of a given definite integral

(9.14)
$$I = \int_X f(x) \, dx ,$$

where the <u>integrand</u> f is a functional, and the <u>interval of integration</u> $X \subset R^n$ are given. It will be assumed that f is Riemann integrable [45], and the problem will be specialized to n = 1; that is, X = [a,b] is an interval on the axis of reals, and f is a real-valued function of a single real variable. The following result, however, holds in general.

Theorem 3.1. If F is an interval extension of the integrand f of (9.14), then

(9.15) $$I \in F(X) \cdot \int_X dx = [c,d].$$

Proof: The upper endpoint of [c,d] is an upper bound for an upper Riemann (or Darboux) sum for the integral (9.14), hence $I \leq d$ [45]. Similarly, c is a lower bound for a lower Riemann sum for the integral, hence $c \leq I$. QED.

Rall [68] has used this simple Riemann-sum type of numerical integration for the numerical solution of integral equations, and the idea has been shown recently to have deeper theoretical extensions and implications [9].

The integration of a single function of a single variable can be considered to be a special case of the integration of a single ordinary differential equation (9.7), to which the method of Moore applies. For

(9.16) $$I'(x) = f(x), \quad I(a) = 0,$$

one has that

(9.17) $$I(b) = \int_a^b f(x)\,dx = I ,$$

the integral to be evaluated numerically. A direct approach to this is to integrate the Taylor polynomial (4.4) for f, and then applying Theorem 3.1 to the remainder term $R_k(b;a)$ by evaluating an interval extension $R_k F([a,b])$ of it:

(9.18) $$I \in \sum_{i=0}^{k} \frac{(b-a)^{i+1}}{i+1} \cdot f_i + \frac{(b-a)^{k+2}}{k+2} \cdot F(k+1) ,$$

where (9.13) has been used. Since the numerical Taylor coefficients $f_i = f_i(a)$ and their multipliers in (9.18) can be calculated using interval arithmetic, as well as the interval Taylor coefficient F(k+1) = F(k+1)([a,b]) and its multiplier, it follows that the inclusion is rigorous if computation is done in this way. Furthermore, if the interval [a,b] is broken up into subintervals of length h = (b - a)/n, then the <u>truncation error term</u>

(9.19) $$E_n = h^{k+2} \sum_{m=1}^{n} (1/(k+2)) \cdot F(k+1)([x_{m-1}, x_m])$$

will be of order $O(h^{k+1})$, and can be made as small as desired if F(k+1)([a,b]) is a bounded interval [27], at the cost of possible increase in the width of the <u>rule</u> of numerical integration,

(9.20)
$$R_n = \sum_{m=1}^{n} \sum_{i=0}^{k} f_i(x_m) h^{i+1} / (i+1) \; ,$$

where $x_m = a + (m-1)h$, $m = 1,2,\ldots,n$.

The numerical coefficients of $f_i(x_m)$ appearing in (9.20) may be simply tabulated; however, the usual approach to numerical integration, as found in standard texts on the subject [18], [49], [58], is to derive formulas for numerical integration which involve only values of the integrand f, and not of its Taylor coefficients (or derivatives). A typical example is Simpson's rule, which expresses the integral (9.17) as

(9.21)
$$I = \int_a^b f(x)\,dx = \frac{b-a}{6}[f(a) + 4f(\frac{a+b}{2}) + f(b)] + \frac{(b-a)^5}{120} f_4(\xi) \; ,$$

where ξ is some (generally unknown) point in the interval [a,b], that is, $a \le \xi \le b$. By taking interval extensions, one has

(9.22)
$$I \in R + E \; ,$$

where the rule is

(9.23)
$$R = \frac{B-A}{6}[F(A) + 4F(\frac{A+B}{2}) + F(B)] \; ,$$

and the error term is, for $H = B - A$,

(9.24)
$$E = \frac{H^5}{120} F(4)([a,b]) \; ,$$

with all indicated operations being performed in interval arithmetic. In (9.24), the letters A, B, H are used to allow for the possibility that a, b, h are not machine numbers, so that one must take small intervals containing them in the actual computation; thus, $a \in A$, where A is exactly representable as a machine interval, etc. In (9.24), automatic differentiation would be used to obtain code to evaluate $F(4)(X)$ for $X = [a,b]$, of course.

Remark 1°. Generally speaking, w(R) will depend only on the roundoff error in the calculation of f at a, $m([a,b]) = (a + b)/2$, and b, as well as roundings in those values and the other numerical computations appearing in (9.23). However, if some of the coefficients of the integrand are specified as intervals, due to uncertainty in the data or the desire to have interval values of the integral which include the results of integrating a whole range of input functions specified by interval coefficients, the effect will also appear in the width w(R) of the interval integration rule R.

Remark 2°. The width w(E) of the interval E is, of course, a measure of the truncation error in the use of the rule R alone as an approximation to the value of I, or, more precisely, of the number

(9.25)
$$s = \frac{b-a}{6}[f(a) + 4f(\frac{a+b}{2}) + f(b)]$$

as an approximation to I, supposing that it could be computed exactly. In the case
of interval computation, however, the interval E also includes the effects of any intervals involved in the definition of the integrand or its fourth Taylor coefficient
(or derivative), and thus w(E) also reflects the spread of values due to imprecise
data. Thus, the formulation (9.22)-(9.24) of an interval-valued numerical integration takes into account all possible sources of error from the data, roundoff in the
actual calculations, and truncation.

Technically speaking, Simpson's rule refers to the use of the number s given by
(9.25) as an approximation to the integral I. The interval Simpson's rule (9.22)-
(9.24), that is,

$$(9.26) \qquad S = \frac{B-A}{6}[F(A) + 4F(\tfrac{A+B}{2}) + F(B)] + \frac{H^5}{120} F(4)([a,b]) ,$$

provides not only a numerical approximation by taking s = m(S) or s = h(S) or some
other point in the interval S, but also automatic, guaranteed error bounds. Furthermore, the interval integration rule allows one to find an interval which contains the
integrals of a whole range of input integrands defined by some interval parameters,
in which case, "the interval is the answer." This kind of calculation can be useful
if one is designing software for an application in which only a certain class of functions is to be integrated numerically, and this must be done with a specified precision.

It is easy to take integration formulas from classical numerical analysis, such
as those found in [18], [49], [58], and transform them into interval form. For example, interval versions of Newton-Cotes formulas include Simpson's rule (9.26), the
interval trapezoidal rule

$$(9.27) \qquad T = \frac{B-A}{2}[F(A) + F(B)] + \frac{H^3}{6} F(2)([a,b]) ,$$

and the interval Newton's three-eights rule

$$(9.28) \qquad N = \frac{B-A}{8}[F(A) + 3F(\tfrac{B+2A}{3}) + 3F(\tfrac{2B+A}{3}) + F(B)] + \frac{H^5}{270}F(4)([a,b]) .$$

A number of interval rules for numerical integration are given in the report
[28], with truncation error terms expressed in terms of derivatives, which can easily
be transformed into the equivalent expressions in terms of interval Taylor coefficients. Interval versions also exist for Gaussian integration rules [51] and Euler-
Maclaurin formulas [29]. Indeed, any rule which involves a linear combination of
values of the integrand and its derivatives or Taylor coefficients at points in the
interval of integration, and a truncation error term based on values of derivatives
or Taylor coefficients at unknown points in the interval of integration (or even outside) can be transformed into an interval rule as done above. Software for calculation with various interval rules for numerical integration has been implemented in
the program INTE [28], which will be described in more detail in the next section.

Remark 3°. In the use of an interval integration rule based on a classical rule for numerical quadrature, such as the interval Simpson's rule, a certain amount of information may be generated which is not used in the calculation. For example, the evaluation of the remainder term (or error term)

(9.29) $$E = \frac{H^5}{120} F(4)([a,b])$$

of (9.26), if done recursively, requires the evaluation of the interval Taylor coefficients

(9.30) $$F(0)([a,b]), \ F(1)([a,b]), \ F(2)([a,b]), \ F(3)([a,b]) \ ,$$

which are used for no other purpose in the calculation. However, these could be used in forming other intervals containing the value I of the integral being computed numerically, for example, by the use of (9.18). Interval computations have what is known as the intersection property, that is, if it is known that $I \in I_1$ and $I \in I_2$, then

(9.31) $$I \in I_1 \cap I_2 = I_3 \ .$$

This means that if several interval evaluations are made of the same quantity (which itself can be an interval), then it is worth saving the intersection of all previous results as I_2 and then using (9.31) with the current result I_1, since the accuracy of I_3 (measured by its width) will be at least as good as any previously computed interval. This prevents loss of information, and can result in considerable improvement in certain cases. In ordinary computing, by contrast, one is not sure but that additional calculations may make the results worse, due to increased round-off or other malignancies.

In classical numerical analysis, considerable ingenuity was required to obtain formulas of high accuracy (measured by remainder terms involving derivatives or Taylor coefficients) by using only a few evaluations of the function, and avoiding differentiation. With the coming of automatic differentiation, there is open the exciting prospect of new developments in numerical analysis in which both ingenuity and derivatives are allowed.

4. The program INTE. The program INTE [28] is essentially a software package which automates the application of various interval rules for numerical integration, such as (and including) (9.26), (9.27), and (9.28). In addition to the original content of this program, described in [27], [28], capability for automatic Euler-Maclaurin integration [29] has been added, in which the rule, as well as the error term, involves derivatives of the integrand. The program also contains some optimization features with respect to the accuracy and total computation time, as will be explained below.

The purpose of the program INTE is not so much to compute numerical integrals themselves as to automate certain aspects of classical numerical analysis. From the output of this program, one obtains not only a value of the numerical integral, but also an error analysis of the result automatically. Thus, as a typical application, this program could be used to choose among several rules of numerical integration to find one which is both economical to implement and accurate enough for a given application. This automates what would be a tedious chore if done manually. In addition, since the error bounds provided are rigorous, this program could be used to tabulate the values of functions defined by definite integrals, or in other situations in which guaranteed error bounds are deemed to be necessary, as in the design of critical components of air or space vehicles. It is to be emphasized, however, that the present version of INTE should be thought of as a research tool, not as a production-type program for numerical integration.

The program INTE operates in batch or interactive mode; in the latter case, a certain amount of explanatory matter is printed to guide the user of the program. Basically, the input consists of an integer L, which gives the number of subintervals into which the interval of integration [X1,X2] is to be subdivided. If the user gives the value L = 0, then a desired accuracy EPS may be specified, and the program will try to choose L so as to attain that accuracy in the integration. (The method for choosing L will be described below.) The next input, N, is a positive integer giving the number of nodes in each subinterval for the rule of numerical integration. This provides a way to increase the accuracy of each rule (Newton-Cotes, Gaussian, etc.) by increasing the order of the corresponding remainder term. The integrand F will be defined as a function of the variable of integration X by a FORTRAN-type expression as described previously. At the option of the user, shorthand variables V1, V2, ... can be introduced before F is defined, and the formulas given can also contain parameters P1, P2, ... The usual order in which to specify input is (1) values of the parameters (optional), (2) formulas for shorthand variables (optional), (3) the formula for the integrand F.

After the above input, the interval rule of numerical integration is chosen from a menu of available rules, to which others can be added if desired. The present version of INTE includes the following rules:

RIEMANN - This simply breaks the subinterval of integration specified by the value of L into N subintervals, and applies Theorem 3.1 to each. The output value for the interval integral I will then be the sum of all the intervals obtained, and will contain the value (interval or real) of the integral being calculated. The value of N is unrestricted.

TRAPEZOIDAL, $2 \leq N \leq 25$. For N = 2, this selects the interval trapezoidal rule (9.27) in each subinterval. For larger values of N, the extended trapezoidal formula [18], [49], [58] is computed in interval form. Output consists of the rule R, the error term E, and the interval integral I = R + E in the chosen interval format.

SIMPSON, N odd, $3 \leq N \leq 25$. For N = 3, this invokes the interval formula (9.26), while the extended Simpson formula [18], [49], [58] is used for larger values of N within the specified range.

NEWTON-COTES CLOSED, $2 \leq N \leq 9$. This specification provides the interval trapezoidal rule (9.27) for N = 2, the interval Simpson's rule (9.26) for N = 3, and the interval Newton's three-eights rule (9.28) for N = 4. For larger values of N, the interval version of the corresponding formulas (see, for example, [49], pp. 123-124) are used for the calculation.

NEWTON-COTES OPEN, $4 \leq N \leq 10$. These are interval versions of the formulas which can be found for example, in [49], pp. 126-127 (the remainder term in formula (1) on p. 127 of [49] should read $3y''h^3/4$), and are useful in the construction of so-called "predictor-corrector" methods for the solution of ordinary differential equations numerically, as exemplified by Milne's method [48], [49], [50]. These open rules omit the values of the integrand at the endpoints A, $B = A + (N - 1)*H$ of the sub-interval of integration, where

$$(9.32) \qquad\qquad H = (X2 - X1)/(L*(N - 1)) \; .$$

For example, for $N = 5$, one has

$$(9.33) \qquad\qquad I = \int_a^b f(x)\,dx \in Z_o \; ,$$

where the interval Newton-Cotes open formula [28] is

$$(9.34) \qquad Z_o = \frac{4H}{3}[2F(A+H) - F(A+2H) + 2F(A+3H)] + \frac{112H^5}{15} F(4)([a,b]) \; .$$

OPEN-CLOSED PAIRS, $N = 4,5,11,13,25$. These interval integration rules are based on Milne's observation [48] that the Newton-Cotes closed and open numerical integration rules can be paired in such a way as to use the same set of nodes in the rule and the same Taylor coefficient of the integrand in the error term, and that while the multiplier of the Taylor coefficient in the open rule is larger than for the corresponding closed rule, it has an opposite sign, so that the error can "cancel out" if the results are "averaged" with the proper weights. The closed rule corresponding to (9.34) ($N = 5$) is Simpson's rule twice, which in interval form is

$$(9.35) \qquad Z_c = \frac{H}{3}[F(A) + 4F(A+H) + 2F(A+2H) + 4F(A+3H) + F(A+4H)] - \frac{8H^5}{15} F(4)([a,b]) \; .$$

Thus, once the closed interval rule (9.35) has been calculated, all of the information necessary to evaluate (9.34) has been obtained, so the evaluation of Z_o is essentially "free". Rall [27], [28] noted that for

$$(9.36) \qquad\qquad I = \int_0^2 \sqrt{1 + 4x} \; dx = \frac{13}{3} \; ,$$

the use of the Simpson formula (9.35) gave (in midpoint $\pm$ halfwidth notation):

$$(9.37) \qquad\qquad I \in Z_c = 4.41392885 \pm 0.08329545 \; ,$$

which is not very accurate, while the open rule (9.34) gave

$$(9.38) \qquad\qquad I \in Z_o = 3.17915720 \pm 1.16613360 \; ,$$

which is quite a bit worse. However, by the intersection principle (9.31), one has

$$(9.39) \qquad\qquad I \in Z_c \cap Z_o = 4.33796210 \pm 0.00732870 \; ,$$

a result which has one more decimal place of accuracy than (9.37). Since the calculation of a suitable open-closed pair of interval integration formulas can be done
very economically once the closed formula is evaluated, and can result in an increase
in accuracy, a set of formulas of this type have been derived, and are available as
an option in INTE (see [28], Appendix A).

GAUSS, $2 \leq N \leq 10$.   The Gaussian integration rules of classical numerical analysis (see [49], pp. 285-288, [18], [58]) are of the form

(9.4)
$$\int_a^b f(x)\,dx = \sum_{i=1}^n w_i f(x_i) + C_n \cdot f(2n)(\xi) \ ,$$

similar to the Newton-Cotes formulas, except that the <u>nodes</u> x_1, x_2, ..., x_n and the
<u>weights</u> w_1, w_2, ..., w_n are transcendental numbers, in contrast to (9.32) and the sim-
ple rational numbers encountered before.   Consequently, interval values of these constants are stored in a table contained in the program INTE for the standard interval
of integration $[-1,1]$, to which the given interval of integration $[a,b]$ is transformed.   (The transformation is done in interval arithmetic, to include the effect of any
roundoff error in the final result.)   Also, the values of C_n (see, for example, [51])
are obtained as intervals containing the true results.   Examination of the error term
in (9.40) reveals that Gaussian integration rules have an order of accuracy about two
times as high as the Newton-Cotes closed rules based on the same number of nodes.

EULER-MACLAURIN, $N \geq 2$, $0 \leq K \leq 9$.   The capability to perform interval Euler-
Maclaurin integration was added to the program INTE subsequent to the publication of
the report [28]; this feature is described in [29].   The basic idea of this approach
to numerical integration is to correct the trapezoidal formula on the basis of Taylor
expansion of the integrand at the endpoints of the interval of integration; hence,
Taylor coefficients of f appear in the integration rule as well as the error term.
In its interval formulation, the Euler-Maclaurin rule is the sum of the extended trapezoidal formula,

(9.41)
$$TN = H * [\frac{F(A) + F(B)}{2} + \sum_{i=1}^{n-1} F(A+iH)] \ ,$$

a <u>correction</u> <u>term</u>

(9.42)
$$SNK = - \sum_{m=1}^{K-1} \frac{H^{2m} * B_{2m}}{2m} * [F(2m-1)(B) - F(2m-1)(A)] \ ,$$

which involves the Taylor coefficients $F(2m-1)$ of F at A and B, and the <u>Bernoulli</u>
<u>numbers</u> B_{2m}, which are rational numbers tabulated in Table IX.1 [58], and, finally,
an <u>error</u> (or <u>remainder</u>) <u>term</u>

(9.43)
$$RNK = - H^{2K} * (B - A) * B_{2K} * F(2K)([a,b]) \ .$$

Thus, the <u>interval</u> <u>Euler-Maclaurin</u> <u>formula</u> with N nodes and order 2K is

(9.44)
$$EMNK = TN + SNK + RNK .$$

TABLE IX.1. THE BERNOULLI NUMBERS $B_2, \dots, B_{18}$

$$B_2 = \frac{1}{6},$$

$$B_4 = -\frac{1}{30},$$

$$B_6 = \frac{1}{42},$$

$$B_8 = -\frac{1}{30},$$

$$B_{10} = \frac{5}{66},$$

$$B_{12} = -\frac{691}{2730},$$

$$B_{14} = \frac{7}{6},$$

$$B_{16} = -\frac{3617}{510},$$

$$B_{18} = \frac{43867}{798}.$$

In addition to the selection of the rule of numerical integration, other communication with INTE is accomplished in the batch version by means of control cards, in which the first 14 characters of the control word appear in columns 1-14. Usually, only the first six columns of a control card are checked. An image of each control card is printed, and descriptive information can be displayed on the control card after a blank column following the control word. In addition to the control words cited above, the following are recognized by INTE:

COMMENT - This control card is simply printed out, and INTE proceeds to the next control card. A card with columns 1-12 blank will accomplish the same purpose.

PRINTOUT - This card must be followed by a data card read according to the 16I5 format, which resets the KODOUT array to modify the printing of the output of the program according to the wishes of the user. The KODOUT array controls the output from INTE in the following manner:

KODOUT(1) $\neq$ 0: Print the intermediate value of the integration rule at each node of integration.

KODOUT(2) $\neq$ 0: Print the value of the function F at each node of integration.

KODOUT(3) $\neq$ 0: Print the value of each node of integration.

KODOUT(4) $\neq$ 0: Print the value of the corresponding weight at each node of integration.

KODOUT(5) $\neq$ 0: Print the value of the error term E after each integration.

KODOUT(6) $\neq$ 0: Print the value of the Taylor coefficient of the integrand appearing in the error term.

KODOUT(7) $\neq$ 0: Print the value of the rule of numerical integration R after each integration.

KODOUT(8) $\neq$ 0: Print the value of the numerical integral $I = R + E$ after each integration.

KODOUT(9) $\neq$ 0: Print the value of the total computation time for each numerical integral I.

KODOUT(10) $\neq$ 0: Print the values of the midpoint and absolute error bound (the half-width).

KODOUT(11) $\neq$ 0: Print the values of the harmonic mean and relative error for each numerical integral I.

KODOUT(12) $\neq$ 0: Print the value of the harmonic mean and percentage error for each numerical integral I.

If the value of any word in the KODOUT array is zero, then the corresponding printout is omitted. All entries in the KODOUT array in the batch version of INTE are set to zero initially.

INITIALIZE - This control card causes the location in the program used for storing the values of the numerical integral, the rule of numerical integration, and the error term to be reset to zero. This will allow the user to recompute the value of the numerical integral without redefining the endpoints, the number of nodes of integration, or the function to be integrated. Prior to the reinitialization, the current value of the numerical integral is stored in another location.

INTERSECTION - This control card causes INTE to calculate the intersection of the current interval value of the numerical integral with the interval value of a previously computed numerical integral which was saved by the use of INITIALIZE.

CONTINUE - This does not clear the locations assigned to the value I of the numerical integral, the rule R of numerical integration, or the error term E. The values of the next calculation will be added to these values. The user may redefine any or all of the endpoints, number of nodes, number of subintervals, integrand, and formula for numerical integration. This instruction is useful in the piecewise computation of an integral

$$(9.45) \qquad I = \int_a^b f(x)\,dx = \int_a^p f(x)\,dx + \int_p^b f(x)\,dx, \quad a < p < b,$$

which can be particularly helpful if the integrand has certain properties or peculiarities which make different methods of numerical integration appropriate in different subintervals of the interval of integration [a,b].

RESET - This card resets INTE to its original state except for printout options. Following a non-fatal error, INTE will ignore control cards until a RESET card is encountered. All input data (except printout options) have to be redefined at this point.

RESTART - This is used in the interactive version of INTE to start computing at the beginning of the program, so that the printout options can be changed, if desired.

CHANGERULE - In the interactive program, this allows for changes in the rule of integration and the endpoints of the interval of integration without requiring re-entry of the formula for the integrand. The information necessary for the given rule, namely, the number of nodes and subdivisions of the interval of integration, must be entered after this instruction is used.

ENDATA - This concludes the input deck and terminates the execution of the program when encountered.

The program INTE also contains a feature which optimizes the computation with

respect to certain criteria. Except for the Euler-Maclaurin integration method, which
depends on the parameters N and K and thus has a more complex structure with respect
to optimization [29], and the Riemann sum procedure, the user has the option of having the program choose the minimum value of L (the number of times the given rule is
applied to the interval of integration) to: (1) attain a given accuracy in terms of
absolute error, or (2) attain the maximum possible accuracy. An estimate of the time
required is made before the calculation is started, to give users in the interactive
mode a chance to decide whether to proceed, or perhaps try a different numerical integration method, or abandon the problem. The optimization is only approximate, but
has proved to be highly effective in actual computation [102].

In order to use the optimization feature of INTE for the integration formulas of
Newton-Cotes and Gaussian types, the user first specifies

$$(9.46) \qquad L = 0 .$$

If a specified accuracy is desired, then option (1) is taken by assigning the desired
limit for $\frac{1}{2} \cdot w(I)$, the half-width of the interval numerical integral as EPS, a double
precision floating point number, for example,

$$(9.47) \qquad EPS = 5.D-5$$

if four decimal places of accuracy are desired. The value

$$(9.48) \qquad EPS = 0.0$$

will elect option (2); the program will try to produce a result with the maximum possible accuracy. If the value of EPS specified in (9.47) is smaller than the accuracy
the program decides is attainable, the value for option (2) will be calculated with a
notice printed to that effect.

In order to make the desired decisions about accuracy, the program applies the
specified rule of numerical integration once to the interval of integration, and obtains the interval value J for the integral being computed in the form

$$(9.49) \qquad J = R + E ,$$

where the error term E is of the form

$$(9.50) \qquad E = C \cdot H^{k+1} ,$$

in which the constant C depends on a Taylor coefficient of the integrand evaluated
over the interval of integration [a,b]. The number

$$(9.51) \qquad r = \frac{1}{2} \cdot w(R)$$

is a measure of the roundoff error in the calculation of the rule of numerical integration, and, of course, any interval component of the value of the integrand due to
interval coefficients, etc. The number r forms a lower bound to the value $\frac{1}{2} \cdot w(I)$ as
the number of applications of the rule of numerical integration is increased; there
is no way that the integral can be computed more accurately by the given formula than

given by (9.51). The input value of EPS will be compared to $\hat{r}$, where

(9.52)
$$\hat{r} = 5 \cdot 10^{-\hat{p}-1} > r ,$$

for the <u>maximum</u> value possible $\hat{p}$; that is, the value which corresponds to $\hat{p}$ decimal places of accuracy, which is the most that can be attained because of (9.51). If EPS $\leq \hat{r}$, then the option corresponding to (9.48) will be taken. Thus, the program sets EPS $= \hat{r}$ if option (2) is selected by (9.48), or if the given value of EPS is less than or equal to $\hat{r}$, and this value is considered to be the smallest possible error which can be attained.

As the number of applications L of the rule of numerical integration is increased, the half-width r of the rule is assumed to remain constant, due to the averaging feature of such rules. This assumption has proved to be adequate for practical purposes, since optimal values of L turn out to be small in most cases. An empirical study of the behavior of $\frac{1}{2} \cdot w(R)$ shows that it does increase slowly, something of the order of $\log(L)$ being observed as L increases, but this variation is much less rapid than the change in the error term, for which the behavior [28] is like:

(9.53)
$$\frac{1}{2} \cdot w(E) = (1/L)^k \cdot \frac{1}{2} \cdot w(C \cdot H^{k+1}) = (1/L)^k \cdot t ,$$

where the initial value of the half-width (9.50) of E is used for the value of t. It follows that one will have

(9.54)
$$EPS \geq \frac{w(R)}{2} + (1/L)^k \cdot t = r + (1/L)^k \cdot t \geq \frac{1}{2} \cdot w(I) ,$$

if L is taken to be the smallest integer satisfying

(9.55)
$$L > (t/(EPS - r))^{1/k} ,$$

assuming that $w(R)/2$ is essentially constant. This value of L is the one taken by the program INTE to optimize the calculation.

As an example, the calculation of

(9.56)
$$I = \int_0^2 [\sqrt{1 + 4x} + \sin(17x)] dx$$

is considered. This example is not as innocent as it looks; the derivatives of the square root have a pole at $x = -1/4$, close to the interval of integration, and the sine is highly oscillatory. Consequently, the Taylor coefficients of the integrand increase rather rapidly. The interval Newton three-eights rule (9.28) will be used for the integration. The values of KODOUT(K) for K = 9, 10, 11 were set equal to 1, with the rest being 0. The remainder of the input was:

        L = 0
        EPS = 5.D-5
        N = 4
        X1 = 0.

```
X2 = 2.

F = SQRT(1.+4.*X)+SIN(17.*X)
```

NEWTON-COTES CLOSED

The output of the initial computation made for the purpose of optimization, and the final result of the optimized calculation were:

```
SUM OF R = ( .32792683+01   /   .32792696+01 )
SUM OF E = (-.4124488+03    /   .41363461+03 )
SUM OF I = (-.40916961+03   /   .41691388+03 )
```

```
L  (OPTIMIZED) = 54
ESTIMATED ERROR = .1465793775-005
ESTIMATED TIME = 3.620 SECONDS
```

```
SUM OF R = ( .44420741+01   /   .44420767+01 )
SUM OF E = (-.12012920-04   /   .68926851-05 )
SUM OF I = ( .44420620+01   /   .44420836+01 )
```

```
THE MIDPOINT IS .444207280+01 A   .000001080+01
THE HARMONIC MEAN IS   .44420728+01 R   .24780377-05

    TOTAL COMPUTATION TIME = 3.782 SECONDS
```

The above example is taken from [28], where the actual computer printout is reproduced. The report [28] also includes a listing of the program INTE on microfiche, which does not include the Euler-Maclaurin subroutine described in [29], which was added to the software at a later date.

An improved program of this type could be produced in which provision is made for input of the formulas for the weights, nodes, and rules of numerical integration, together with the error term, all expressed as linear combinations of values of the Taylor coefficients (or derivatives) of the integrand, all of which would be computed automatically by the program. As indicated by Rall [74], programs of this type could also be used for automatic error analysis and optimization of formulas from classical numerical analysis for interpolation and numerical differentiation, since the truncation error terms are normally expressed in terms of derivatives of the function being dealt with approximately.

5. Projects for Chapter 9.

1°. Write software to automate the generation of the Taylor series expansion of a function given (1) by a formula; (2) by an initial-value problem of the form (9.7)-(9.8).

2°. Write software for the calculation of the Taylor polynomial and remainder term in interval arithmetic, thus obtaining automatic error estimates. Assume that TYPE INTERVAL is available.

3°. Write a version of the program INTE which, instead of having a fixed menu of formulas for numerical integration, will accept a subroutine for the rule of numerical integration, and a formula for the error term, expressed in terms of derivatives or Taylor coefficients of the integrand. Apply the result, for example, to automate the error analysis of Weddle's rule ([49], p. 125),

$$(9.57) \qquad \int_{x_0}^{x_6} f(x)\,dx = \frac{3h}{10}(f_0 + 5f_1 + f_2 + 6f_3 + f_4 + 5f_5 + f_6) - \frac{h^7 f^{(6)}}{140} - \frac{9h^9 f^{(8)}}{1400} \ ,$$

where $f_i = f(x_i) = f(x_0 + ih)$, $i = 1,2,\ldots,6$.

4°. Write software to automate the error analysis of standard formulas for polynomial interpolation ([49], p. 76) and numerical differentiation ([49], pp. 96-99).

CHAPTER X

ADDITIONAL NOTES ON TECHNIQUES, APPLICATIONS, AND SOFTWARE

> "Though for no other cause, yet for this,
> that posterity may know that we have not
> loosely through silence permitted things to
> pass away as in a dream."
>
> *RICHARD HOOKER*

The previous chapters have been concerned mainly with techniques for automatic differentiation, together with applications and software developed at the Mathematics Research Center. Although the techniques are of general applicability, the restriction of attention to specific programs has been dictated by personal familiarity and the fact the software described actually works. There have undoubtedly been many programs of real value developed elsewhere for automatic differentiation and the generation of Taylor coefficients. Unfortunately, it is possible that a lot of this work has wound up in the ashcan due to being tied too closely to outmoded machines, operating systems, or languages, or because of personnel shifts, administrative decisions (not necessarily wise), and so on. The purpose of this chapter is to give some historical perspective to the subject, and to mention some techniques, applications, and software developed elsewhere, for which references are available. It is realized that the discussion below is by no means exhaustive, so apologies are tendered in advance to those who have labored long and hard to add this capability to their software, but are not mentioned.

1. Generation of Taylor coefficients. The use of Taylor series in numerical analysis has a long history, and was well-developed as early as 1730, when the remarkable book by James Stirling [85] appeared. In fact, it is probably fair to say that classical numerical analysis is based on computation with polynomial approximations to functions, and the Taylor polynomial in particular. The idea of recursive generation of Taylor coefficients is also not new. In the survey paper by Barton, Willers, and Zahar [2], references are made to the use by Airey in 1932 and J. C. P. Miller in 1946 and in the National Physical Laboratory Tables of 1954-5 [46], [47], of the set of recurrence relation formulas for Taylor coefficients. The use of full recurrence is attributed to J. C. P. Miller in connection with the preparation of the tables [47] in 1954. Also, the paper [2] contains the following remark:

"The method of Taylor series is conceptually straightforward, yet mathematically elegant. Its use has been restricted and its numerical theory neglected merely because adequate software in the form of automatic programs for the method has been nonexistent."

The paper [2] cited above was published in 1971. Software for the generation of Taylor series, however, has existed at least since 1964 [57], as described in Chapter 4. Unfortunately, media for the publication of computer programs was not well developed at the time, so that a lot of software described only in technical reports and proprietary documents was delayed in coming to light, and some may still be obscure.

Many of the formulas of classical and modern numerical analysis, particularly for the numerical solution of differential equations, are designed to use function evaluations in such a way as to be "as good as" Taylor series in the sense that they are as accurate as the Taylor polynomial of a certain degree. Given software for the automatic generation of Taylor coefficients, it might be just as well to use them directly, especially since estimates of truncation error can also be obtained automatically by interval evaluation of the remainder term expressed in terms of the next highest Taylor coefficient.

In connection with the use of Taylor series, mention can also be made of the Lie series method [99], which is essentially a perturbation procedure. In the program described in [100], the Taylor polynomial was used as an approximate solution of a system of differential equations, and then the Lie series computation was applied to obtain a more accurate solution. The user, however, had to analyze the formulas defining the system of equations, and write the corresponding sequence of calls to subroutines in order to obtain the Taylor coefficients. A coder was added to the proram [100] by Julia Gray and Tom Szymanski to obtain the completely automatic program [101].

2. <u>Straightforward differentiators</u>. Programs for differentiation of formulas also appeared early in the history of large-scale machine computation. There is a reference in [33] to work done in the Soviet Union as early as 1959 [3]. The paper [33] by Hanson, Caviness, and Joseph, was published in 1962, and describes a sophisticated coder/differentiator. The goal of this program, however, was to compress the output into a formula for the derivative, using lots of parentheses, in a way similar to the corresponding feature in SUPER-CODEX. The authors do remark that the program could just as well produce machine code, which is the object of the differentiators described in this book. It is perhaps worthy of note that [33] was published in the "Unusual Applications" section of the Communications of the ACM, showing that a community which had already accepted automatic formula evaluation was not yet ready for automatic differentiation.

Other early papers on automatic differentiation include [88] and [91]. In [88], however, the process of translation of the formula into a sequence of calls to subroutines (the function of the coder), is presumed to be done by hand, and thus cannot be considered to be truly automatic.

A more modern automatic differentiator is the program <u>pdgen</u> of D. D. Warner [87]. The abstract of the report [87] describing this program reads:

"The necessity for solving systems of nonlinear equations arises in many contexts.

A case of particular importance is the solution of stiff systems of ordinary differential equations. The fastest numerical methods for solving systems of nonlinear equations revolve around Newton's Method and at some stage require a subroutine for computing the Jacobian Matrix, i.e. the matrix of partial derivatives. Writing this subroutine is typically a straightforward but tedious and error-prone task. The program, pdgen, reads the definition of the system of nonlinear equations, symbolically computes the partial derivatives of the component functions, and generates a FORTRAN subroutine which will evaluate the component functions and the corresponding matrix of partial derivatives. The system of nonlinear equations is defined using a FORTRAN-like grammar. Pdgen is capable of differentiating arbitrary algebraic expressions and expressions involving the following elementary functions: SQRT, EXP, LN, LOG10, COS, SIN, TAN, ARCCOS, ARCSIN, ARCTAN, COSH, SINH, TANH, ARCCOSH, ARCSINH, and ARCTANH. In addition, pdgen has a feature whereby the user can introduce differentiation rules for arbitrary functions.

"The program pdgen is written in C and runs on both the UNIX and GCOS systems at Bell Laboratories."

The capabilities and goals of pdgen thus resemble the ones of the programs described previously, CODEX, SUPER-CODEX, and NEWTON. The feature of interval arithmetic available in the latter programs could be added to the FORTRAN programs obtained as output from pdgen by use of the AUGMENT precompiler, assuming that the necessary arithmetic modules were available for the computer being used.

3. Symbolic algebraic manipulators. The discussion up to now has been limited to software which will analyze formulas and produce code for evaluation of derivatives and Taylor coefficients of the functions considered. Software (and software systems) of the symbolic manipulation category have the much more extensive objective of performing much of the tedious work involved with the manipulation of polynomials, algebraic expressions, and formulas in general (including partial sums of infinite series, and even explicit integration), in an automatic fashion. A number of these systems include differentiation as an available operation. The field is very extensive at the present date; for an introduction, see [8], which includes a useful bibliography of 52 items.

Two of the many symbolic manipulators are FORMAC [86] and ALTRAN [6]. According to [6], ALTRAN had its origins in 1964, and is the successor to an earlier system called ALPAK, which dates back to about 1962. A very useful reference for ALTRAN and its applications is the bibliography [7], which lists 68 references in the following categories: Primary references (4), Background (5), Theory and implementation (14), Applications of ALPAK (1963-65) (10), Applications to SIGSAM problems (6), Applications to algebraic coding theory (9), Miscellaneous applications (14), and Surveys (6). Needless to say, the field is far beyond the scope of this book.

Another class of software of this kind is more special-purpose in nature. For example, the optimization program PROSE [84] and other developments of optimization language [64] depend on imbedded differentiators, see also [103], pp. 27-30.

REFERENCES

1. Airey, J. R.: Emden Functions. British Association for the Advancement of Science Mathematical Tables, Vol. II.B.A., London, 1932.
2. Barton, D., Willers, I. M., and Zahar, R. V. M.: Taylor series methods for ordinary differential equations - an evaluation, [81], pp. 369-390 (1971).
3. Beda, L. M., Korolev, L. N., Sukkikh, N. V., and Frolova, T. S.: Programs for automatic differentiation for the machine BESM (Russian). Institute for Precise Mechanics and Computation Techniques, Academy of Science, U.S.S.R., Moscow, 1959.
4. Binstock, W., Hawkes, J., and Hsu, N.-T.: An interval input/output package for the UNIVAC 1108. MRC Tech. Summary Rept. No. 1212, University of Wisconsin-Madison, 1973.
5. Braun, J. A. and Moore, R. E.: A program for the solution of differential equations using interval arithmetic (DIFEQ) for the CDC 3600 and 1604, MRC Tech. Summary Rept. No. 901, University of Wisconsin-Madison, 1968.
6. Brown, W. S.: ALTRAN User's Manual, 4th Ed., with contributions from S. I. Feldman, S. C. Johnson, and A. D. Hall. Bell Laboratories, Murray Hill, N. J., 1977.
7. Brown, W. S.: ALTRAN Bibliography. Bell Laboratories, Murray Hill, N. J., 1978.
8. Brown, W. S. and Hearn, A. C.: Applications of symbolic algebraic computation. Computer Physics Communications $\underline{17}$ (1979), 207-215.
9. Caprani, O., Madsen, K., and Rall, L. B.: Integration of interval functions. SIAM J. Math. Anal. $\underline{12}$ (1981) (to appear); MRC Tech. Summary Rept. No. 2087, University of Wisconsin-Madison, 1980.
10. Conte, S. D.: Elementary Numerical Analysis: An Algorithmic Approach. McGraw-Hill, New York, 1965.
11. Conte, S. D. and de Boor, C.: Elementary Numerical Analysis: An Algorithmic Approach, 2d Ed. McGraw-Hill, New York, 1972.
12. Crary, F. D.: Language extensions and precompilers. MRC Tech. Summary Rept. No. 1319, University of Wisconsin-Madison, 1973.
13. Crary, F. D.: The AUGMENT precompiler. I. User information. MRC Tech. Summary Rept. No. 1469, University of Wisconsin-Madison, 1974. Revised, 1976.
14. Crary, F. D.: The AUGMENT precompiler. II. Technical documentation. MRC Tech. Summary Rept. No. 1470, University of Wisconsin-Madison, 1975.
15. Crary, F. D.: A versatile precompiler for nonstandard arithmetics. ACM Trans. Math. Software $\underline{5}$, 2 (1979), 204-217.
16. Crary, F. D. and Ladner, T. D.: A simple method of adding a new data type to FORTRAN. MRC Tech. Summary Rept. No. 1605, University of Wisconsin-Madison, 1970.
17. Davidenko, D. F.: On a new method for the solution of systems of equations (Russian). Dokl. Akad. Nauk SSSR $\underline{88}$ (1953), 601-602.
18. Davis, P. J. and Rabinowitz, P.: Numerical Integration. Blaisdell, Waltham, Mass., 1967.
19. Davis, P. J. and Rabinowitz, P.: Methods of Numerical Integration. Academic Press, New York, 1975.
20. Dennis, J. E. Jr.: Toward a theory of convergence of Newton-like methods, [72], pp. 425-472 (1971).
21. Dennis, J. E. Jr. and Schnabel, R. B.: Quasi-Newton Methods for Unconstrained Nonlinear Problems, Lecture Notes, Rice University, Houston, 1979.
22. Foster, J. M.: List Processing. American Elsevier, New York, 1967.
23. Forsythe, G. E. and Moler, C. B.: Computer Solution of Linear Algebraic Systems. Prentice-Hall, Englewood Cliffs, N. J., 1967.
24. Gragg, W. B. and Tapia, R. A.: Optimal error bounds for the Newton-Kantorovich method. SIAM J. Numer. Anal. $\underline{11}$ (1974), 10-13.
25. Gray, Julia H. and Rall, L. B.: NEWTON: A general purpose program for solving nonlinear systems. MRC Tech. Summary Rept. No. 790, University of Wisconsin-Madison, 1967.
26. Gray, Julia H. and Rall, L. B.: NEWTON: A general purpose program for solving nonlinear systems. Proceedings of the 1967 Army Numerical Analysis Conference,

U. S. Army Research Office, Research Triangle Park, N. C., 1967, pp. 11-59.

27. Gray, Julia H. and Rall, L. B.: A computational system for numerical integration with rigorous error estimation. Proceedings of the 1974 Army Numerical Analysis Conference, U. S. Army Research Office, Research Triangle Park, N. C., 1974, pp. 341-355.

28. Gray, Julia H. and Rall, L. B.: INTE: A UNIVAC 1108/1110 program for numerical integration with rigorous error estimation. MRC Tech. Summary Rept. No. 1428, University of Wisconsin-Madison, 1975.

29. Gray, Julia H. and Rall, L. B.: Automatic Euler-Maclaurin integration. Proceedings of the 1976 Army Numerical Analysis and Computers Conference, U. S. Army Research Office, Research Triangle Park, N. C., 1976, pp. 431-444.

30. Gray, Julia H. and Reiter, A.: Compiler of differentiable expressions (CODEX) for the CDC 3600. MRC Tech. Summary Rept. No. 791, University of Wisconsin-Madison, 1967.

31. Hansen, E. R.: Interval arithmetic in matrix computation. SIAM J. Numer. Anal. $\underline{2}$ (1965), 308-320.

32. Hansen, E. R. and Sengupta, S.: Global constrained optimization using interval analysis, [60], pp. 25-47 (1980).

33. Hanson, J. W., Caviness, J. S., and Joseph, C.: Analytic differentiation by computer. Communications ACM $\underline{5}$ (1962), 349-355.

34. Hassitt, A.: Design and implementation of a general-purpose input routine. Communications ACM $\underline{7}$ (1964), 350-355.

35. Kantorovich, L. V.: Functional analysis and applied mathematics. Uspehi Mat. Nauk $\underline{3}$ (1948), 89-185. Tr. from Russian by C. D. Benster, Natl. Bureau of Standards Rept. No. 1509, U. S. Dept. of Commerce, Washington, D. C., 1952.

36. Kantorovich, L. V.: On a mathematical symbolism convenient for performing machine calculations (Russian). Dokl. Akad. Nauk SSSR $\underline{113}$ (1957), 738-741.

37. Kedem, G.: Automatic differentiation of computer programs. ACM Trans. Math. Software $\underline{6}$, 2 (1980), 150-165.

38. Krawczyk, R.: Newton-Algorithmen zur Bestimmung von Nullstellen mit Fehlerschranken. Computing $\underline{4}$ (1969), 187-201.

39. Kuba, D. and Rall, L. B.: A UNIVAC 1108 program for obtaining rigorous error estimates for approximate solutions of systems of equations. MRC Tech. Summary Rept. No. 1168, University of Wisconsin-Madison, 1972.

40. Li, T.-Y. and Yorke, J. A.: A simple, reliable numerical algorithm for following homotopy paths, [82], pp. 73-91 (1980).

41. Mancini, L. J. and McCormick, G. P.: Bounding global minima. Math. Operations Res. $\underline{1}$ (1976), 50-53.

42. Mancini, L. J. and McCormick, G. P.: Bounding global minima with interval arithmetic. Operations Res. $\underline{27}$ (1979), 743-754.

43. McCarthy, J. et al.: LISP I Programmers Manual. Computing Center and Research Laboratory, Massachusetts Institute of Technology, Cambridge, Mass., 1960.

44. McCormick, G. P.: Computability of global solutions to factorable nonconvex programs: Part I - Convex underestimating problems. Tech. Paper Serial T-307, Inst. for Management Sci. and Engr., George Washington University, Washington, D. C., 1975.

45. McShane, E. J.: Integration. Princeton University Press, Princeton, N. J., 1944.

46. Miller, J. C. P.: The Airy Integral. British Association for the Advancement of Science Mathematical Tables, Part-Vol. B, Cambridge University Press, London, 1946.

47. Miller, J. C. P.: Introduction to Tables of Weber Parabolic Cylinder Functions. National Physical Laboratory, H.M.S.O., London, 1955.

48. Milne, W. E.: Numerical integration of ordinary differential equations. Amer. Math. Monthly $\underline{33}$ (1926), 455-460.

49. Milne, W. E.: Numerical Calculus. Princeton University Press, Princeton, N. J., 1949.

50. Milne, W. E.: Numerical Solution of Ordinary Differential Equations. Wiley, New York, 1953. Reprinted by Dover, New York, 1970.

51. Moore, R. E.: The automatic analysis and control of error in digital computation based on the use of interval numbers, [66], pp. 61-130 (1965).

52. Moore, R. E.: Automatic local coordinate transformations to reduce the growth

of error bounds in interval computation of solutions of ordinary differential equations, [67], pp. 103-140 (1965).

53. Moore, R. E.: Interval Analysis. Prentice-Hall, Englewood Cliffs, N. J., 1966.

54. Moore, R. E.: A test for existence of solutions to nonlinear systems. SIAM J. Numer. Anal. 14 (1977), 611-615.

55. Moore, R. E.: Methods and Applications of Interval Analysis. SIAM Studies in Applied Mathematics, 2, Society for Industrial and Applied Mathematics, Philadelphia, 1979.

56. Moore, R. E. and Jones, S. T.: Safe starting regions for iterative methods, SIAM J. Numer. Anal. 14 (1977), 1051-1065.

57. Moore, R. E., Davison, J. A., Jaschke, H. R., and Shayer, S.: DIFEQ integration routine - User's manual. Tech. Rept. LMSC 6-90-64-6, Lockheed Missiles and Space Co., Palo Alto, Calif., 1964.

58. Mysovskih, I. P.: Lectures on Numerical Methods. Tr. from Russian by L. B. Rall, Wolters-Noordhoff, Groningen, The Netherlands, 1969.

59. Nickel, K.: On the Newton method in interval analysis. MRC Tech. Summary Rept. No. 1136, University of Wisconsin-Madison, 1971.

60. Nickel, K. (Ed.): Interval Mathematics 1980. Academic Press, New York, 1980.

61. Ortega, J. M.: The Newton-Kantorovich theorem. Amer. Math. Monthly 75 (1968), 658-660.

62. Ortega, J. M. and Rheinboldt, W.: Iterative Solution of Equations in Several Variables. Academic Press, New York, 1970.

63. Ostrowski, A.: Solution of Equations and Systems of Equations. Academic Press, New York, 1960. Second edition, 1966; Third edition, 1973.

64. Pugh, R. E.: A language for nonlinear programming problems. Math. Program. 2, 2 (1972), 176-206.

65. Rabinowitz, P. (Ed.): Numerical Methods for Nonlinear Algebraic Equations. Gordon and Breach, New York, 1970.

66. Rall, L. B. (Ed.): Error in Digital Computation, Vol. 1. Wiley, New York, 1965.

67. Rall, L. B. (Ed.): Error in Digital Computation, Vol. 2. Wiley, New York, 1965.

68. Rall, L. B.: Numerical integration and the solution of integral equations by the use of Riemann sums. SIAM Rev. 7 (1965), 55-64.

69. Rall, L. B.: Solution of abstract polynomial equations by iterative methods. MRC Tech. Summary Rept. No. 892, University of Wisconsin-Madison, 1968.

70. Rall, L. B.: Davidenko's method for the solution of nonlinear operator equations. MRC Tech. Summary Rept. No. 948, University of Wisconsin-Madison, 1968.

71. Rall, L. B.: Computational Solution of Nonlinear Operator Equations. Wiley, New York, 1969. Reprinted by Krieger, Huntington, N. Y., 1979.

72. Rall, L. B. (Ed.): Nonlinear Functional Analysis and Applications. Academic Press, New York, 1971.

73. Rall, L. B.: A comparison of the existence theorems of Kantorovich and Moore. SIAM J. Numer. Anal. 17 (1980), 148-161.

74. Rall, L. B.: Applications of software for automatic differentiation in numerical computation. Computing, Suppl. 2 (1980), 141-156.

75. Rall, L. B.: Representation of intervals and optimal error bounds. MRC Tech. Summary Rept. No. 2098, University of Wisconsin-Madison, 1980.

76. Reiter, A.: Compiler of differential expressions (CODEX). Prog. No. 1, Mathematics Research Center, University of Wisconsin-Madison, 1965.

77. Reiter, A.: Interval arithmetic package (INTERVAL). Prog. No. 2, Mathematics Research Center, University of Wisconsin-Madison, 1965.

78. Reiter, A.: Automatic generation of Taylor coefficients (TAYLOR). Prog. No. 3, Mathematics Research Center, University of Wisconsin-Madison, 1965.

79. Reiter, A.: Interval arithmetic package (INTERVAL) for the CDC 1604 and CDC 3600. MRC Tech. Summary Rept. No. 794, University of Wisconsin-Madison, 1967.

80. Reiter, A.: Automatic generation of Taylor coefficients (TAYLOR) for the CDC 1604. MRC Tech. Summary Rept. No. 830, University of Wisconsin-Madison, 1967.

81. Rice, J. R. (Ed.): Mathematical Software. Academic Press, New York, 1971.

82. Robinson, S. M. (Ed.): Analysis and Computation of Fixed Points. Academic Press, New York, 1980.

83. Sokolnikoff, E. S. and Sokolnikoff, I. S.: Higher Mathematics for Engineers

and Physicists. McGraw-Hill, New York, 1945.

84. Stark, R. L.: PROSE General Information Manual. PROSE, Inc., Palos Verdes Estates, Calif., 1980.

85. Stirling, J.: Methodus Differentialis: sive Tractatus de Summatione et Interpolatione Serierum Infinitarum. Typis, Gul. Bowyer, Inpensis, G. Strahan, London, 1730.

86. Tobey, R. G. et al.: FORMAC. SHARE Contributed Program Library, No. 360, D-0.3.3004, IBM, White Plains, N. Y., 1969.

87. Warner, D. D.: A partial derivative generator. Comp. Sci. Tech. Rept. No. 28, Bell Laboratories, Murray Hill, N. J., 1975.

88. Wengert, R. E.: A simple automatic derivative evaluation program. Communications ACM 7 (1964), 463-464.

89. Wertz, H. J.: SUPER-CODEX (Supervisor plus a compiler of differentiable expressions). Mathematics Research Center, University of Wisconsin-Madison, 1968.

90. Wertz, H. H.: SUPER-CODEX: Analytic differentiation of FORTRAN statements. Rept. No. TOR-0172 (9320) - 12, Aerospace Corporation, El Segundo, Calif., 1972.

91. Wilkins, R. D.: Investigation of a new analytical method for numerical derivative evaluation. Communications ACM 7 (1964), 465-471.

92. Yohe, J. M.: The interval arithmetic package. MRC Tech. Summary Rept. No. 1755, University of Wisconsin-Madison, 1977.

93. Yohe, J. M.: Implementing nonstandard arithmetics. SIAM Rev. 21 (1979), 34-56.

94. Yohe, J. M.: Portable software for interval arithmetic. Computing, Suppl. 2 (1980), 211-229.

Additional reference on automatic differentiation:

95. Kedem, G.: Automatic differentiation of computer programs, MRC Tech. Summary Rept. No. 1697, University of Wisconsin-Madison, 1976.

Additional references on automatic error analysis:

96. Bauer, F. L.: Computational graphs and rounding error. SIAM J. Numer. Anal. 11 (1974), 87-96.

97. Stoutemeyer, D. R.: Automatic error analysis using computer algebraic manipulation. ACM Trans. on Math. Software 3 (1977), 26-43.

98. Stummel, F.: Rounding error analysis of numerical algorithms. Computing, Suppl. 2 (1980), 169-195.

Additional references on Lie series:

99. Knapp, H. and Wanner, G.: Numerical solution of ordinary differential equations by Groebner's method of Lie-series. MRC Tech. Summary Rept. No. 880, University of Wisconsin-Madison, 1968.

100. Knapp, H. and Wanner, G.: LIESE: A program for ordinary differential equations using Lie-series. MRC Tech. Summary Rept. No. 881, University of Wisconsin-Madison, 1968.

101. Knapp, H. and Wanner, G.: LIESE II - A program for ordinary differential equations using Lie-series. MRC Tech. Summary Rept. No. 1008, University of Wisconsin-Madison, 1969. '

Additional reference on numerical integration:

102. Rall, L. B.: Optimization of interval computation, [60], pp. 489-498 (1980).

Additional reference on software:

103. Knott, G.: MLAB, An On-line Modeling Laboratory, Reference Manual, 8th Ed., Div. of Computer Research and Technology, Natl. Inst. of Health, Bethesda, Maryland, 1979.

SUBJECT INDEX